Improving Vehicle Workshop Efficiency

Operators and Users Committee

J Ashmore
Trent and Barton Buses

J Baker
Ryder Trucks

D Bass
Automobile Association

G Boughton
Parcelforce

D Brain
Chassis Developments Limited

C Copelin
Confederation of Passenger Transport UK

D Cox
Powerfleet Services

G Day
Freight Transport Association

R Denniss (Committee Chairman)
Lorry logic

C Evans
Volvo Bus and Truck

D Hay
DETR

P Howard
BT Group Motor Transport

S Hunter
Automobile Association

S Nancollis
ERF Limited

N Samuel
Glass's Information Services Limited

J York
National Freight Corporation

Organizing Committee Members

Gordon Spooner
Manager – Vehicle Maintenance
Metropolitan Police Service

Alan Parker
Director of Engineering
TNT UK Limited

Papers presented at a one-day seminar *Improving Vehicle Workshop Efficiency – Incorporating Automatic Identification Technology* held at the Gas Research and Technology Centre, Loughborough, UK, on 18 November 1998.

IMechE
Seminar Publication

I MECH E

Improving Vehicle Workshop Efficiency
Incorporating Automatic Identification Technology

Organized by the Operators and Users Committee of the Automobile Division of the Institution of Mechanical Engineers (IMechE)

Co-sponsored by

IRTE
GEA
The Leasing Group (a subsidiary of BG plc)
Transco

IMechE Seminar Publication 1999–5

Professional Engineering Publishing

Published by Professional Engineering Publishing Limited for the Institution of Mechanical Engineers, Bury St Edmunds and London, UK.

First Published 1999

ISSN 1357–9193
ISBN 1 86058 228 1

A CIP catalogue record for this book is available from the British Library.

Printed and bound in Great Britain by Bookcraft (Bath) Limited.

Contents

Related Titles of Interest

Title	Editor/Author	ISBN
Automotive Braking – Recent Developments and Future Trends	David Barton	1 86058 131 5
Handbook of Automotive Body and Systems Design	John Fenton	1 86058 067 X
Handbook of Automotive Powertrain and Chassis Design	John Fenton	1 86058 075 0
Handbook of Automotive Body Construction	John Fenton	1 86058 073 4
Vehicle Handling Dynamics	John R Ellis	0 85298 885 0

For the full range of titles published by Professional Engineering Publishing contact:

Sales Department
Professional Engineering Publishing Limited
Northgate Avenue
Bury St Edmunds
Suffolk
IP32 6BW
UK

Tel: 01284 724384
Fax: 01284 718692

S626/001/98

Maintaining progress

A PARKER
TNT UK Limited, Atherstone, UK

INTRODUCTION.

The scope of this conference focuses on the efficient and cost effective management of a vehicle maintenance workshop. However, before we hear the many answers to the problems facing the transport industry, let me try and outline some of the "big ticket" issues which our particular profession has to accommodate.

Over recent years, there has been a rapidly changing structure of industry and trade which generates freight transport demands. The range and quality of the final products is increasing and the proportion of movements of traditional bulk products is decreasing. Global sourcing is changing the scale and structure of world trade flows, with an increased amount of cross trading in manufacturers at various stages of the fabrication and assemble process.

"Just in time" logistics systems demand more frequent and reliable movements of goods in smaller consignments, often over longer distances with hierarchically centralised warehousing systems. This has emerged from a combination of improved technology for batch processing, falling real freight transport costs, improved control of flows through telematics, and high interest rates. Logistics service is increasingly provided on external contract to specialists as manufacturing companies retreat to their core business.

Another area of significant change has been vehicle design and manufacturing process. Improved styling and attention to the vehicles aerodynamic properties has provided tangible benefits in fuel consumption and importantly, costs. Driveline efficiencies have been improved together with tyre performance all of which should add value to the operator, driver and ultimately the customer.

MAKING MONEY.

Like most of you attending today, TNT is very much focused on making money and importantly, making a profit at an acceptable margin. An increasing cost base coupled with a very competitive market, means that all cost areas are not only dynamically examined but benchmarked against others in the industry. One sometime forgotten area of cost are the maintenance systems employed either inside or outside the operators total control.

The annual cost of maintaining a typical four axle tractor unit and trailer combination for instance, is in excess of £8,000. At 20%, to some of you this may not seem a large figure when compared to the total operating cost of that combination. However, if the maintenance systems are not thought out properly, implemented and managed in line with demands, not only can the costs escalate beyond the bounds of acceptability, but adverse legal and safety implications are very real possibilities. In addition, if the maintenance is not undertaken correctly, other areas of costs can be significantly and adversely effected - fuel economy, availability, vehicle performance.

MINIMISING COSTS.

Assuming that, the operational demands have been correctly investigated and more importantly understood, the correct equipment specification has been calculated and the right procurement mechanism has been chosen, you then need to decide which one of the many methods of maintaining this valuable asset should be used over its life, in order to ensure that through maximum availability, the vehicle performs at the lowest operating cost. How do you do this ? Listen to the following speakers and read their papers thoroughly.

Some organisations are turning to external service providers to undertake these activities and deliver a business solution to meet their needs. These organisations, however, are broadly using the term "outsourcing" to encompass a wide range of contract services. A more accurate term might be "out-tasking". One particular area which cannot be "outsourced" is responsibilities and there are some organisations who are not fully aware of this. Likewise, some maintenance providers either do not fully understand or use this perceived ignorance to gain competitive advantage.

CUSTOMER DEMANDS.

The customer I am referring to in this section is you, the operator of the vehicle. You are demanding increased availability, reduced running costs, zero roadside failures, minimal front end costs and maximum residuals. A tall order ?

As a service provider operating within the framework of a P+L, my colleagues and I fully understand these demands and the need for you to derive a quality margin out of your revenue streams. We also appreciate that your customer demands are getting more difficult to achieve, margins are getting increasingly smaller and road congestion is getting worse.

The development of our infrastructure appears to be stuck on "pause" and your basic utilitarian vehicle is being re-designed into something resembling the control panel of the "Star Ship Enterprise". All of which you perceive as an on-cost with little value and a cost which you are finding increasingly more difficult to justify and pass on to your customer.

We also appreciate your environmental pressures. Notwithstanding the majority source of pollutants - private transport rather than freight movements - it's the lorry which has received the biggest criticism, and against the use of which there are increasing calls for restrictions. In particular, three innovations in logistics have led to changes in the nature of the transport element and, hence have had an effect on the environment :

- rationalisation of production and distribution leading to fewer inventory-holding facilities, this leads to greater transport requirements between centres of production and consumption.

- 24-hour deliveries, which may be more fuel efficient during night-time driving and lead to less urban congestion, but may have a negative impact on the quality of sleep of people living on the delivery routes.

- just in time deliveries, which lead to more frequent distribution of smaller loads, minimising storage and therefore a greater number of journeys. In turn, this leads to an increase in "transport-related" environmental problems.

THE MARKET.

Our market, which is the vehicle maintenance market, is becoming more and more confusing to say the least. Subject to who you're talking to, and on which day, there is a difference in opinion as to what the market needs are. In addition, I believe that those who are in the industry have lost their way insofar as what is their core business.

I hear of vehicle manufacturers and their dealers who, as a result of minimal margins on the sale of a truck, have begun to experiment with not only the provision of a vehicle on contract hire, but to include a driver. Perhaps the next phase is for them to address the whole logistical needs of their customers - interesting thoughts ?

From my perspective, the trend in the market place appears to be one of :

- vehicle manufacturers and their dealer networks offering a range of down-stream services which I suspect that they do not fully understand.
- contract hire companies becoming the middle men between the manufacturer and operator.
- little or no training initiatives resulting in a skills shortage within the industry.
- once employed, little or no appreciation showed resulting in poor retention levels.
- vehicle technology being heavily marketed, sometimes for the wrong reasons.

Contents

- Introduction
- Making Money
- Minimising Costs
- Customer Demands
- The Market
- Conclusion

Introduction

- Changing structure of industry and trade
- Just in time logistics systems
- Vehicle Design and manufacturing process

Making Money

- Tractor unit & trailer :
    - maintenance costs in excess of £8,000 per annum
    - total operating cost - in excess of £40,000 per annum
- Maintenance as a percentage of total - 20%

Making Money

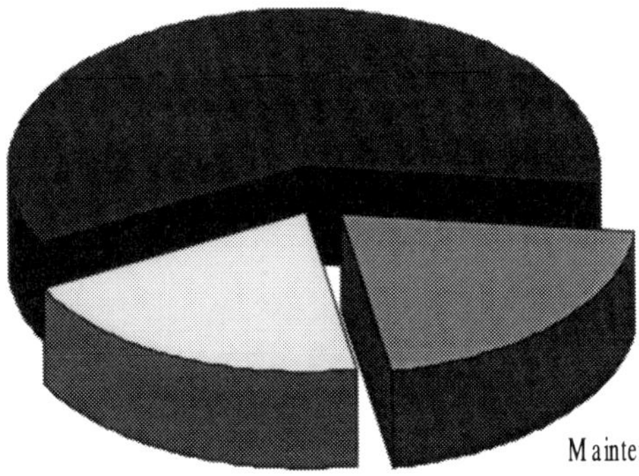

Note : life - 5 years + 5 years for the trailer, GTW - 32 tonne, mileage - 100,000 miles p/annum

Minimising Costs

- Operational demands
- Vehicle specification
- Procurement mechanism
- Maintenance disciplines and procedures
  - in-house
  - package deal
  - outsourcing to an external service provider
- Legal responsibilities

Customer Demands

- Rationalisation of production and distribution
- 24-hour deliveries
- Just in time

The Market

- Manufacturers and dealers offering down-stream services
- Contract hire companies - middle men
- Training initiatives
- Staff retention
- Vehicle technologies

S626/002/98

Workshops – friend or foe

T MINETT
TLS Contract Hire, Wednesbury, UK

SYNOPSIS

Written from the perspective of a major Vehicle Rental and Contract Hire operator, the paper seeks to overview the requirements placed on such an operator by its customers, and hence the key elements of a modern service required from a vehicle maintenance/workshop operator.

CUSTOMER DEMAND

Vehicle Rental and Contract Hire customers have become progressively more demanding. These demands fundamentally focus on the following areas:-

1 The need to perform the same operations with fewer vehicles - this leads to significant pressure to increase vehicle availability.

2 Significant cost pressure not only on vehicle rentals but particularly on recharge costs - the understanding and traceability/consistency of such costs.

3 A requirement for a higher and more bespoke level of service - including mobile/flexible services and dedicated contract management.

4 Many large operators have centralised transport management with a small staff interfacing with one or a few transport partners. This has lead to the re-emergence of partnerships between vehicle providers and major customers and the increasing use of open book contract management.

The above approach places a significant emphasis on the provision of detailed and wide ranging management information, particularly relating to:-

1 Vehicle breakdown - time to attend/repair.

2 Vehicle downtime - routine servicing.

3 Out of contract charges - understand consistency.

4 Total compliance with legal servicing including document availability.

WORKSHOP SUPPLY

Faced with such customer demands Vehicle Rental and Contract Hire providers are faced with a significant gap in terms of the service provided by many workshops.

The key areas which will be developed in the presentation are as outlined:-

Vehicle Availability
Although the requirement for 100% operational vehicle availability is apparently understood, out of hours and overnight servicing are still difficult to obtain in many areas. Many overnight servicing facilities are restricted in terms of the services offered, particularly delivery and collection. The use of mobile fitters is still relatively rare amongst workshops and the use of mobiles can significantly improve vehicle availability.

Cost Pressures
There is a requirement for workshops to agree standard tariff rates for the replacement of regular recharge items.

Times booked and parts prices often vary significantly leading to customer non payment or conflict with the customer.

Service Levels/Responsiveness
The presentation will detail required response times for major contracts and show typical roadside and repair responses which are being achieved.

Management Information
The production of accurate management reports detailing sometimes thousands of transactions per month requires precise administration and additional information from workshops.

Examples of downtime reporting and attendance response reports are shown in Figure 1.

Workshops are often unable to meet these management information requirements or adopt sufficiently rigorous operational standards to allow Vehicle Rental and Contract Hire companies to reliably use in house IT to record, for example -attendance time.

The presentation will further outline management information required from workshops, and show further examples of the management information produced for Contract Hire customers.

The report will also outline the success rate of workshops in completing the administration of PMI and MOT documentation within a suitable timeframe.

THE FUTURE

As Rental and Contract Hire companies draw closer into partnerships with their customers with increasing service and information requirements, it is likely that Vehicle Rental and Contract Hire companies will form partnerships with maintenance providers. Such a relationship is likely to be based on the same criteria of reduced administration and enhanced service, which has provided the impetus for the partnerships between Vehicle Rental and Contract Hire companies and their own customer base.

This will lead to significant linkages in the areas of IT and significantly enhanced customer management by maintenance providers.

SEPTEMBER 1998

Vehicle Reg	Division	Date (In)	Location	Model	Details ?		Time (hrs)	Call Out	Attend
S809 SKE	B.T.	29/08/98	DUDDEN HILL	IVECO 26T T/G	ENGINE OIL LEAK	**	1.50	13.54	14.25
N919 XJH	B.T.	04/09/98	DUDDEN HILL	IVECO 75E15	DOUBLE BLOW OUT ON O/S/R	**	ABORTED		
S809 SKE	B.T.	08/09/98	DUDDEN HILL	IVECO 26T T/G	CRANE OIL LEAK		4.00	9.35	10.00
N918 XJH	B.T.	09/09/98	DUDDEN HILL	IVECO 14FT FLAT	AIR LEAK		1.50		
R554 CUG	B.T.	09/09/98	DUDDEN HILL	IVECO 35-10	LOSS OF BRAKES		22.00	8.10	10.30
R141 AVM	B.T.	10/09/98	DUDDEN HILL	IVECO 140E15	INDICATORS U/S	**	1.50		
R557 CUG	B.T.	10/09/98	DUDDEN HILL	IVECO 35-10	REAR LIGHT LENS	**	0.50		
R917 LKN	B.T.	11/09/98	DUDDEN HILL	35-10	NO ELECTRICS		2.00		
M823 NMO	B.T.	14/09/98	DUDDEN HILL	TRANSIT 190	EXHAUST/TOW PLATE		5.00		
R873 JAD	B.T.	14/09/98	DUDDEN HILL	L/DAF 32T	BEACONS/CLUTCH		1.25		
R584 EHU	B.T.	17/09/98	DUDDEN HILL	L/DAF 26T	CRANE SLOW		1.25		
R584 EHU	B.T.	17/09/98	DUDDEN HILL	L/DAF T/G	GRAB STUCK IN AIR	**	5.50	7.55	8.50
R554 CUG	B.T.	21/09/98	DUDDEN HILL	IVECO 35-10	BRAKES FAILED		27.00	7.55	9.30
R972 UKK	B.T.	22/09/98	DUDDEN HILL	IVECO 75E15	STEERING DAMAGED		22.00		
R772 CJO	B.T.	23/09/98	DUDDEN HILL	ASTRA D VAN	WINDSCREEN		2.00	10.21	12.00
S809 SKE	B.T.	25/09/98	DUDDEN HILL	IVECO 26T T/G	NO GEARS	**	0.50	8.51	9.25
S809 SKE	B.T.	26/09/98	DUDDEN HILL	IVECO 26T T/G	ACCELERATOR INOP	**	0.50	11.26	12.00

Shows % of Down Time in relation to Normal Hours
** = Jobs handled by RACCA

S626/003/98

Assessing and improving workshop efficiency – 'the tangibles'

C J SUMMERSELL
Motor Vehicle Repairers' Association, Blackburn, UK

Creating an efficient workshop environment is certainly a thought provoking ideal which requires the correlation of an awful lot of information in order to commence the process. The important factors which need to be taken into consideration are many and varied, and careful thought and planning are required to optimise the facilities that exist already, and which also apply even if you are fortunate enough to have a new large site, a clean sheet of paper and a blank cheque.

I read the other day that bodyshop planning could be regarded as an art form, so with that in mind let's give thought to how we could start this creative process.

The following points are put forward in the order in which I think they should be addressed – you may disagree, but they all have to be covered in my opinion. It is important to realise that although they may appear to be about a new facility, which I am sure you will agree is an ideal situation, every one of them can be considered and applied to an existing workshop. The reason for this is that although changes could be radical ones, very often they are subtle, based perhaps on just looking at one area which possibly could be improved by looking at a 'perfect' scenario, and seeing if that could be replicated in an existing facility, thereby helping to achieve some of the efficiencies that we are discussing today.

Let us remember that we are trying to achieve the highest volume of work at the lowest possible cost, and that must always, in whatever decisions we take, be uppermost in our thought process. Let us look at the possible first step in creating this perfect scenario.

1. What size and shape of building is, or needs to be, available.

How long is a piece of string? If the building is already built, to some extent you have a 'fait accompli'. However, it still needs looking at very carefully as it is not always first decisions, which may have been snap ones, that are correct, and discussion can very often open one's eyes to a rejigging of an existing facility with a cost effective improvement programme.

If you are planning a new building it is easy, or is it? This starts with all that has been said already, plus what size and shape and where do we place it on the site? I have seen many new workshops where insufficient consideration has been given to the placement of the building on the site.

Which leads neatly on to

2. What ratio of vehicle storage or parking to workshop capacity is required?

This is a crucial factor in the efficient running of any workshop. All too often I see highly paid personnel spending time moving vehicles, either to gain access to other vehicles or to make room to park up or move vehicles into the workshop. The cost involved per annum in this type of exercise can run into thousands of pounds in lost productive time, and time spent planning the areas and maximising storage and creating easy vehicle movement can pay huge dividends.

I saw an example recently where the sweep of a pair of gates took up twelve parking spaces in an area where currently you would only be able to park approximately twenty cars. A set of sliding gates cured the problem!

Other areas to consider in this scenario are damage. When parking is not disciplined space can become tight and damage to vehicles and property soon occurs. How often have we seen the gateposts bent because there is not enough turning area from the road to allow sensible access? The result one damaged post and one damaged vehicle!

The other area that has recently received attention in the press is the increase in reversing accidents, not only causing damage to property but more importantly injury, sometimes very serious, to staff and even customers.

Safety rails inside perimeter fences allowing safe clearance when vehicles are parked is a good safety measure. Entrances and exits should be of the correct size, and positioned such that access both in and out of workshop and parking areas can be achieved with little or no disturbance to other vehicles or work.

3 What categories of vehicle or plant are to be serviced?

Any operation should always be geared around those vehicles or plant for which it is required to provide a service.

If the fleet is light, for example car and the lighter end of the light commercial range, size is not so important, as in general terms job times are short and very varied and so a lot of space is not always necessary, unless you have a large fleet.

However, if the requirement is for heavy goods vehicles and the heavy end of the light commercial market, then size becomes very much an issue, not just because of the size of vehicles involved, but also because the work carried out takes longer and in many cases, such as pre MOT preparation, would mean that one bay could be tied up for as much as seven working days. Thus vehicle turnaround is slower, and where you are dealing with vehicles that are not earning money if they are idle problems can soon arise! The more efficient you are the less downtime is created, and excess downtime, as we all know, is a sure way to raise cost or lose profitability whichever way you view it!

Bus and coach operation has similar requirements to HGV but working hours tend to differ as a lot of this work has traditionally been done at night.

Plant is probably the other category that needs some extra thought, as a cleaning and fabrication facility will have to be included on top of the service requirement for this category.

So we have already started our thought process and can now move on.

Equipment is our next step. Layout must offer flexibility and be correctly positioned so as to ensure that throughput and efficiency are maintained whatever job is currently in progress. Even if you have existing equipment installed, could it be re-sited, or possibly updated, in order to create more work area and/or better work flow?

One of the most useful tools we have to help us in this endeavour is the use of computer aided design systems. These enable us to look at so many of the variables involved with the siting of major pieces of equipment before we actually do it!

We must now look at what other equipment we may require in our operation.

Fixed equipment and plant are reasonably easy to source and site, but there is still a lot more to be considered, especially in terms of tooling which can cover trolley jacks, welding equipment, service aids and tools, and just about any other small item that is purchased! Every item of equipment requires evaluation to see if its purchase will aid throughput efficiency and safety. We need to ask the question 'does it not only fill our current requirements but also cover a possible increase in usage over the next ten years?'

This question would apply especially to fixed installations and plant, however, small tooling is just as important and should not be ignored. Some tools are 'trendy' but actually have a very short useful life, but even if that is the case they may still make a valuable contribution during that period. There are also many tried and tested pieces of equipment that we can comfortably purchase with confidence, but we should still carry out the evaluation process to ensure that mistakes are avoided. The following questions should be asked - 'do we need it, will it be fully utilised, and what benefits in terms of efficiency will we gain?'

With the current plethora of equipment on the market, whatever is chosen must be cost-effective, not only in terms of price but serviceability and backup in the event of breakdowns. Equipment that does not function for any reason at all when it is required to do so is of no use whatsoever!!

Lighting and heating are other areas that need consideration. Let's look at lighting first. It is probably more important than we realise this because it tends to be something we take for granted. First of all good lighting should provide sufficient illumination of the workplace. Together with the power supply the installation should be safe and well maintained, and it should also be of the correct type for whatever work area it is installed in. Good lighting means that hazards can be identified quickly, mistakes are more easily avoided and staff do not end up having time off work for eyestrain, migraine and headaches!

But still in workshops we find poor bench lighting, poor pit and ramp lighting, uneven lighting of work areas creating areas of shadow, lights poorly positioned so as to dazzle workers, the list goes on. Remember, lighting helps our ability to see and any lighting installation should be chosen with the following criteria in mind:

- suitability for use
- safety of use
- size of area to be covered the purpose for which it is used

There should be emergency lighting systems in use to enable staff to vacate the premises safely if required to do so in an emergency. Lamps that give off a stroboscopic effect should not be used to illuminate moving machinery. Where solvent vapours are present e.g. in paint mixing areas and pits, there should be sealed lighting systems with explosive proof switches and fittings. The benefits of good lighting are so important in terms of a reduction in accidents, mistakes and poor heath within the workforce.

Heating a workshop can be a nightmare, both in terms of costs and efficiency. Apart from considering what type of heating to install, thought must be given to the insulation of the building as heating the outside air is a very costly exercise! Many new shops have electronically controlled doors which because of their ease of operation are kept closed for a longer period of time than their manual counterpart. If windows are present, they should be double glazed if possible. The prevention of heat escaping is always a very good exercise to carry out. In terms of deciding what heating system and fuel to use, the cheapest is very rarely the most efficient and some systems that purport to be cheap to install may be just that, but not cheap when it comes to efficiency and running costs. I have worked in many a workshop where the fitter nearest the heater is comfortable and most other staff are freezing cold! There have been other instances of staff being made ill through poorly maintained or installed heating.

At this time we should also consider ventilation, as in hot weather it is as important to have a well ventilated shop as it is to have a warm one in winter.

Parts departments and storage areas should also form part of the planning process. A well but not over stocked, efficient, centrally situated parts department goes hand in glove with achieving efficiencies and cost savings in the workshop as a whole.

Let us now take a look at the legislative requirements of the modern workshop. These are many and varied. I think at the last count there were twenty three sets of regulations which the modern workshop is expected to conform to. The following are probably the most important in the planning process:

1. Pressure Systems and Transportable Gas Containers Regulations 1989
2. Noise at Work Regulations 1989
3. Electricity at Work Regulations 1989
4. Manual Handling Operations Regulations 1992
5. Control of Substances Hazardous to Health Regulations 1994
6. Provision and Use of Work Equipment Regulations 1992

All of these are now mandatory pieces of legislation, and within their structure are contained most of the occupational health standards currently required.

The other most important piece of legislation is obviously The Health and Safety at Work Act 1974 and all that is contained within it. Although we may feel sometimes that we are becoming somewhat bogged down by legislation, I can assure you that, in the final analysis, it is easier and cheaper to conform than ignore, as many businesses are finding to their cost.

This is especially true in the area of staff welfare, accidents and occupational health problems, as litigation in this society that we are part of can be very costly indeed. Good staff facilities are an investment in the future of any company. A happy workforce will always outperform an unhappy one, that is a fact. So some of the tasks we are asked to carry out, such as risk assessments, are not quite as unnecessary as they sometimes seem.

We must also take into account environmental policies, not only those that deal with disposal of waste in all its forms whether water, oil, solvent, or metals etc, but also the formulation of our own policies, as increasingly many of those people who we deal with on a day to day basis are asking for a copy of our policy before they will do business with us! Without a doubt our environment must be looked after and cared for, and as an industry our record in the past has on many occasions been called into question, and to be honest many times justifiably so! Failure to comply is expensive, clean-up operations and in severe cases total stoppage of work are not going to help our efficiency at all, and for many reasons could have a long term effect on our ability to trade successfully in the future.

Finally, a plea from the heart - whatever facility you create, re-design or improve, make sure you have a good housekeeping policy. This is an area where more people and businesses become inefficient than any other, as poorly maintained equipment, piles of rubbish, floors that resemble skating rinks or swimming pools, staff in filthy overalls - the list goes on - spoil our opportunities to work efficiently and gain a good return on our considerable investment at the earliest opportunity.

The dictionary gives us the definition of 'tangible' as 'that can be touched, real'. I like that last word because in the final analysis, the achievement of creating an efficient, comfortable, cost effective workshop is every engineer's idea of utopia, even though, as we see, getting to that stage is a complex process.

"Remember"

Plan carefully
Evaluate fully
Act promptly

S626/004/98

Staff development – the key to workshop efficiency

D TAYLOR
ANC Limited, Newcastle-under-Lyme, UK

Ladies and gentlemen, I'd like to talk to you today about the vehicle workshop's biggest single asset. I'm not referring to the expensive fabric of the building or the computer aided diagnostic equipment of which we are all so proud. I'm talking about the workshop's human resource - **its people** - the apprentices, fitters, chargehands, supervisors, team leaders, foremen and managers, whose collective responsibility it is to keep our vehicles and those of our customers on the road.

As in any other service industry, the primary function of a workshop is to buy and sell man hours and the level of profit made on this sale of labour determines the success of the operation. The profitability of the workshop is directly linked to the quality of its workforce and this is determined not only by the standard of recruitment and basic training, but also on the quality of supervision, appraisal and on-going staff development procedures in place.

Unfortunately the road transport industry is falling down at the very first hurdle - that of recruitment. Whether we like it or not, we have the fundamental problem of not being able to recruit enough young people of the right calibre to train as vehicle mechanics. The workshop is perceived by the computer literate school and college leavers of today, as a dirty, low tech environment in which, as mechanics, they'd have to spend their working day lying underneath a vehicle dressed in greasy overalls and smelling of Derv. It simply doesn't appeal to them as a viable career opportunity and it seems that no one at school or within the careers advisory service is doing much to persuade them otherwise.

This is nothing new, back in the late 70's when I was recruiting apprentice mechanics in the Greater Manchester area, out of the ten school leaving candidates sent along to see me by the careers office, only one had the right qualifications and aptitude for the job. I rang the

Education Department to ask why they had sent us youngsters who were clearly unsuited to the position, only to be told - 'Well, it's only an apprentice mechanics job, we didn't think there would be a minimum requirement'. Why did the Careers Officer think that? Was it his fault or ours? Whichever it was, this was a misconception that prevailed at the time and I suspect, indeed I know, it still lingers today.

Unfortunately, in the eyes of potential employees and their careers advisors, our industry hasn't changed greatly since the 70's. Mechanical engineering is still only recommended to those who are 'good with their hands' and not to those who are academically bright and electronically adept. Why? - because we, as an industry, haven't bothered to inform educators and students of the tremendous technological advances which have been made in the field of vehicle design and maintenance, nor have we advised them of the tremendous employment opportunities on offer in an increasingly high tech and computer orientated working environment.

Enough of the downside. What do we have to do to address the issue?

We need to forge greater links with schools, colleges and careers services to provide advice and guidance on the types of skills which would benefit young people coming into the workplace and through talks, demonstrations and site visits, encourage teenagers to think of vehicle engineering as a highly skilled, dynamic profession worthy of their future consideration. If you haven't got a relationship with your local comprehensive and sixth form colleges, call them. Talk to the Careers Department, I guarantee they will be delighted you've taken the time to contact them and you will be doing yourself and our industry a big favour.

Once we've attracted the youngsters into the industry, we owe it to them to provide the finest level of training possible. If we recognise and respect their importance as the engineers of tomorrow, they will come to value the position and endeavour to reach the high standards we set them. High expectation will be rewarded with high performance. It's definitely a two way process - we'll get back what we invest - and with interest.

Apprentice training should consist of three principal elements; learning one's craft from skilled mechanics on-the-job, college and industry training board programmes of study and courses developed and run by vehicle manufacturers. The three are interlinked and depend upon one another to provide a fully comprehensive training regime.

There's no substitute for real hands-on experience, so don't just let trainees sweep up and make the tea, set them real tasks that give them a sense of achievement.

College and training board programmes provide the essential theoretical component, whilst the involvement of the vehicle manufacturer is crucial in demonstrating how the theory is put into practice. We shouldn't expect our mechanics to repair something they don't fully understand and the manufacturer is best placed to explain how and why the vehicle is put together as it is, the role of each component within the whole and the logic behind the timing of its replacement.

Regular dialogue between colleges, manufacturers and the workplace should continue throughout the apprenticeship and beyond, to ensure that the impact of any new developments, changes in legislation and the latest research findings are communicated, not only to the students but to all the staff.

Training no longer stops when an apprentice has served his time and acquired his City & Guilds or GNVQ certificates. Staff development and re-training is now part of our culture and continues throughout an engineer's working life. Yes it costs money, but what's the alternative? If you think education is expensive, ask yourself - at what price ignorance?

It's possible that a mechanic frozen in a time capsule in 1958 and awakened in 1978, might just have been able to return to his job without too many serious problems. But that certainly wouldn't have been the case for someone sleeping through from 1978 to 1998. Technological advances over the past two decades have meant that the job has changed beyond all recognition. Engineers have had to keep abreast of all the latest developments in vehicle design, diagnostic techniques and maintenance procedures. And from where I'm standing, the rate of change doesn't appear to be slowing down, if anything it's increasing on a daily basis! Our industry is going through a period of rapid evolution and we must be aware of change, embrace change and benefit from it.

In addition to re-training staff to help them keep on top of their present jobs, we should also be actively preparing people for promotion. It's inefficient to wait until a vacancy arises and then hurriedly train someone to fill the role. A professional workshop must be pro-active and have a pool of people trained and ready to step into vacancies as they arise.

If people are absent due to illness or holidays, we should never ask people to cover the jobs below them. It is far more sensible to give people on a lower rung of the promotional ladder, the opportunity to temporarily take on a more senior role. Not only does this give people a taste of what the next job up involves, it also allows you to measure their aptitude to do that job and their ability to cope with the increased responsibility.

Workshops should have a clearly defined hierarchy so that each person can see how their job fits into the overall structure and also identify the promotional routes open to them. Every member of staff must be given a written job description, clearly defining the duties for which they are paid. And they must do the job for which they are paid, not the 'one below' with which they may be more comfortable. A happy workshop is an efficient workshop and letting people know exactly where they stand eliminates discontent and prevents petty squabbles over job demarcation.

A recent study undertaken by Sheffield University advocated keeping workers happy by varying their roles to eliminate boredom. Scania however, have found that their workers don't want variety at the expense of bonus payments - and at the behest of the workforce the company is switching back to 'one worker - one job' so that productivity is raised and bonus payments increased. I suppose you could argue that the real mundane work is now carried out by robots anyway!!

With regard to organisation and supervision, I am firmly convinced that a chargehand should not have responsibility for more than 7-8 fitters. If you give your first line managers too many people to oversee, the system will become 'bottom-heavy' with people wasting time as they wait for guidance and direction. Each chargehand should be responsible for his own team, to stop the mechanics gravitating to one chargehand, usually the weakest.

Fitters' non-productive time, for whatever reason, is a drain on the efficiency and therefore profitability of the business. Strictly speaking, supervisory time may not be 'productive' work, but it is vital to efficiency and must be charged to the job.

Labour is a valuable commodity, but it has a shorter shelf life than perishables like bread and milk. Once you have wasted man hours, you can't get them back. If people are under utilised, then that resource has cost the business money with no return. As the old saying goes - time is money - and it's the workshop manager's job to optimise the use of this precious human asset. If he has 180 hours of labour available and only 160 hours of work, he must go out and sell it like any other commodity and make every hour productive and profitable.

But maximum productivity will only be achieved if you match the right resource to the right job. We often find that it's the best mechanic who's singled out for promotion to chargehand and then foreman. By doing this, we don't always take into account the different skills and attributes required to undertake a supervisory role.

It doesn't necessarily follow that the best mechanic will make the best chargehand; he may have the technical ability but not the supervisory or managerial skills required. The skill difference is like a knife edge and we must help our workers to make this difficult transition by providing adequate training.

I once asked my financial director to send one of his accountants to change a fan belt - he looked at me in amazement. But what's the difference between that and him asking one of my skilled mechanics to read a balance sheet - its exactly the same thing. People aren't born managers. They must be taught how to control budgets, how to promote sales and how to manage a workforce.

If you want someone to do something outside their speciality, you have to provide the additional training. You also have to recognise the strengths of an individual and don't take them off work at which they excel, to do something they don't enjoy and cannot master. If you don't train people to do the jobs you have promoted them to, they will almost certainly fail to reach their full potential.

On the other side of the coin, it is crucial not to sidestep poor workers into other areas where you think they can do less damage to the efficiency of the operation, but in fact, without training, they can do more harm. In the past, how many failed mechanics have been put behind the parts counter with a purchase order book and phone and told to get on with it? If they can't succeed in the job they were trained to do, they haven't a hope of making something of a job they know nothing about.

Auditing efficiency by conducting regular staff assessments is vital. You have to identify the qualities and behaviours you want from your staff to meet your business objectives and then measure progress against these standards. It is essential that people don't view these assessments with fear and dread. It must be explained to them that it is not a disciplinary measure, but one which will benefit them as much as the company. If it's found that their work doesn't meet the required standard, they've probably not received adequate training in that particular facet of the job. It allows you to detect problems and rectify those defects before they have an adverse effect on the operation as a whole.

The importance of this evaluation process is not restricted to the shop floor. By auditing the performance of everyone, from fitters through to senior managers, we can ensure that like the little Dutch boy with his trusty finger in the dam, we can plug any small skill gaps before a minor problem undermines the overall success of the operation.

Workshop staff now need a much broader range of awareness skills than those limited to their own sphere of operation. By being aware of the company's aims and objectives, its position within the industry as a whole and the needs of its customers; individuals will realise how their own job impacts on others within the organisation and understand how their performance and behaviour can make a difference to the efficiency and profitability of the business and the quality of service it provides.

A workshop issue which came to my attention recently, illustrates how a fitter's lack of understanding of the customer's needs almost had catastrophic repercussions. Several vehicles which had failed their MOT were returned to service for their brakes to bed in, prior to the retest. The fitter had mistakenly thought that this was acceptable because the vehicles had been in service for less than a year - he hadn't realised the failure certificate ruled and to operate these vehicles contravened the customer's Operator's Licence. His action could have cost the customer his business.

Our individual actions do matter and they do count. They can affect the profitability of the workshop, they can affect our customers' livelihood and to take the point to the extreme, they can also cost lives. The issue of road accidents resulting from workshop negligence is a real problem - and one we can't hide from.

A quarter of a million people are now involved in vehicle maintenance - this equates to 1.8% of the UK workforce and is falling. As legislation increases our responsibility to our workmates, our customers, the public in general and our environment, who in their right mind would want to come into this industry in the future?

The answer is no-one unless we take some positive action **now**. We must go out and sell our industry and attract the calibre of people who will further its development and not shy away from its significant level of responsibilities. A highly skilled and effective workforce does not come cheap - but in the long term, it will be money very well spent.

To sum up, my recipe for success in a workshop has only got five simple ingredients and these are - to have a well organised structure, to buy and sell labour efficiently achieving high productivity levels and maximum utilisation, resulting in overall sales efficiency.

Materials - to buy materials of the best quality at an acceptable price for own fleet workshops and in retail workshops selling these materials and achieving budgeted margins.

To train people to achieve performances expected in the job the individual is employed to do. This goes a long way to achieving job satisfaction.

To audit to ensure all goals and targets set for the business are being achieved whether they be financial or individual performances.

S626/006/98

Ensuring efficient, cost-effective, and legal maintenance standards

G BOUGHTON
Parcelforce Worldwide, Milton Keynes, UK

Parcelforce operate some 8000 vehicles of varying size and type from Motorcycles through to 38 tonne tractors and trailers. These vehicles mainly operate throughout the UK on Parcel Distribution and Local Delivery work, handling in excess of three quarters of a million parcels each working day, although, an increasing number are engaged on international operations.

The majority of our parcel traffic is carried by road and air, however, a recent development in June 1998, was the introduction of a piggyback operation, (road to rail conversion), from London to Glasgow, with four trailers going in each direction every night.

The majority of vehicles are maintained in house, through a National network of sixteen own account Road Transport Workshops; there are also a larger number of 'satellite' workshops located within delivery depots. These are aligned to the main sixteen workshops and carry out day to day inspection and servicing, with a repair capability which is limited only by space and equipment provided.

However, as a benchmark, and where it is sensible to do so, some 10% of Parcelforce Worldwide vehicles are maintained on contract, by a sister Business Royal Mail Vehicle Services, and some by External Contractors.

The total annual mileage of the fleet is 160 million miles, and some 26 million pounds is spent on fuel.

There are a number of conflicts which arise from operating varying types and sizes of vehicles in this type of environment.

- How for example, do we improve workshop efficiency and maintain quality, whilst reducing the cost of ownership and vehicle operation; using own account workshops maintaining a mixed fleet with a variety of types and models ?

- What systems are necessary to ensure that legal standards are met and sustained, that the maintenance policy fulfils both legal requirements, and yet fully supports the operational requirement ?

The purpose of this presentation is to answer these questions and demonstrate the Parcelforce Worldwide approach to ensuring these criteria are met.

Firstly, why in house?

The only justification for in house maintenance, is reduced cost and greatly improved customer service to the operator.

If internal costs are higher, these will affect the overall profitability of the 'core' business.

If internal customer satisfaction is low, then this will have an effect on the 'core' business's ability to provide and sustain a satisfactory Quality of service to its customers.

To ensure these criteria are met, and to ensure that we can sustain continuous improvement, we must be able to effectively measure our workshop output and performance.

Ensuring vehicles are operated legally and safely are not options, they are the minimum standards. Operating vehicles efficiently in terms of vehicle performance is directly related to effective maintenance standards. After all, the negotiating element on diesel fuel is 1 to 2ppl. One extra mile per gallon, on an average motive unit in Parcelforce Worldwide is £2800/yr, with 450 motive units on fleet, the benefits of regular and effective maintenance are obvious. In this example a loss on average of 1mpg equates to £1.2m on-cost in diesel fuel.

There must also be a balance of cost against quality. It is widely suggested that there is no cost on quality, because a job done right first time, does not result in an early failure, that is, with a breakdown, with the added unnecessary cost this incurs. However, it can be easy to simply strip out cost and ignore effective maintenance at your peril, the only option is to measure performance and to provide adequate resource, whilst aiming to reduce costs through increased productivity and efficiency.

Maintenance Policy.

The basis for effective maintenance must be a robust maintenance policy, then adhering to the policy and ensuring that the standards required are sustained.

The Parcelforce Worldwide approach uses a 48 week maintenance period, utilising internationally recognised week numbers. This leaves 4 maintenance free weeks, one in each quarter, these are aligned to periods of low activity, Bank Holidays for example.

Based on the vehicle type, size, and usage, (either due to mileage or arduous duty) each vehicle is scheduled for a specific number of inspections per annum, these are always capable of being divided equally into 48. For example, 4wks 12 inspections/annum, 6wks, 8 inspections/annum, etc.

The initial planning of the individual vehicle programme is always taken from the annual test as the basis, with inspection weeks being counted either side of this, based on the number of weeks scheduled i.e. 4, 6, etc.

The programme is currently a paper based 'board' system, and is a rolling programme year on year, this ensures that all vehicles receive a set scheduled number of inspections per year, and none are 'dropped off' the schedule, as opposed to some systems where the next inspection is rescheduled from the last inspection date.

Fleet Management System.

The paper system is currently under review to enable electronic scheduling based on our fleet management system PROFLEET, the module already exists within the current system.

An important part of the maintenance programme are the manufacturers recommended Service items, these are aligned to the nearest inspection due, based on either time or mileage, whichever is recommended. The inspection week frequencies are close enough to allow for all items to be scheduled as near as possible to the manufacturers recommendations.

The PROFLEET system produces jobsheets where all activity is recorded, hours, parts, and a record of the activities performed. Through a coding system, overlaid on (VMRS), it is possible to record detail of work done to a high level of detail.

Workshop Management and Technicians can interrogate the PROFLEET system on the vehicle entering the workshop, and can see what has been previously done, how accurate the projected mileage is, and whether some reprogramming of serviceable items either more or less frequently is required. Warranty and all vehicle Modifications can be accessed and recorded.

Whilst we do have our own specific inspection sheets, we have avoided reproducing the manufacturers service sheets, as these are subject to more change than the inspection items. We can add to our inspection sheets, specific items we wish to be checked on a more regular basis in the light of our own operational experience.

Quality control checks are made on individual Technicians work by supervisory staff within the workshop, these are recorded as such on the system. The number of quality control checks conducted can then be interrogated by Territory and Fleet Engineers whilst conducting Technical Audits on workshops. This forms the basis for quality control, and, ensuring that the maintenance Policy and standards are both complied with.

Our workshop based fleet management system as previously referred to, PROFLEET, is a real time system, and a Technician entering time onto a vehicle now, can be accessed within a few minutes of him entering the detail.

PROFLEET is also used for a number of ancillary, but important, fleet management roles, but, is a workshop based system for vehicle maintenance, this is it's primary function.

All workshop staff attendance time is recorded on PROFLEET, both direct time, (time booked directly to vehicles), and Indirect time (supervision, stores control, sick leave, annual leave etc.).

The vehicle asset register is maintained within PROFLEET.

Vehicle jobsheets are created in PROFLEET, and the work detail, with parts and labour hours entered.

Parts are ordered using an order generation module with an uplift into our management accounting system.

Parts stock control and reconciliation is made via PROFLEET.

As would be expected, there are reports available which may be generated on any, or a combination of the above, as well as the facility to construct Ad-Hoc or specialist reports. These can be quite detailed, with several criteria based on either a group of, or individual combinations of the following.

- Depot reports
- Vehicle reports
- Staff reports

PROFLEET feeds basic information into economic life and whole life cost spreadsheets and now has several years of information which can be readily accessed.

The system is currently being upgraded to take account of Year 2000 implications, and the opportunity to make several minor, but important enhancements are being made.

PROFLEET also provides the basic information for our workshop productivity and performance measures and is the basis for all of our fleet related reporting.

Workshop Efficiency.

In order to compare workshop efficiency we must understand that there has to be a relationship between the size of the workshop and the mix of vehicle type and size. For many years, the approach within the Post Office Group has been to use a weighted vehicle factor to compare the repair and maintenance needs of different models and types of vehicle, this is referred to as an 'Equated Vehicle value' or EVV.

This forms the basis for the majority of Parcelforce Worldwide's measurement of workshop performance.

It is worth spending some time on explanation of how equated vehicle values were developed, and, how they have evolved since their introduction during the 1960's.

An equated vehicle value of 1, is directly related to the amount of time required to maintain and repair a base vehicle.

It was derived firstly, by analysing nationally, the number of hours spent on the repair and maintenance of this particular vehicle type, and includes an element of accident and abuse. One equated vehicle value is equal to 37.5 hrs.

The equated vehicle values for all other vehicle models on fleet at that time were then developed from this, and is the basis for all other equated vehicle values for whatever type or model of vehicle is introduced into the fleet.

For example a Tandem axle semi trailer has an equated value of 1.4, whilst a Motive Unit has a value of 3.6. These values, it is important to understand, are model related as each vehicle model has differing maintenance requirements.

The values have not remained stationary, and, only form the basis for measurement.

For example, Motive Unit equated values some years ago were around 4.6. As newer technology or enhancements to vehicles and service intervals have been introduced, this has been reflected in a reduction in the number of hours required to maintain them. The equated vehicle value has therefore been reduced to compensate for higher specification, and or, investment. Some examples of these changes in specification are; Chassis Autolube, Brake system Air dryers, extended oil drain periods etc. etc.

It can be seen from the above that effort is made to recover investment in higher specification by reduced maintenance hours.

What we measure - How and against what?

In setting up a workshop from new, we would use a planning figure of 50 equated vehicles per man. This will include all supervision, stores, direct and indirect hours.

This is only part of the story, but is the basis for our measurement of productivity.

We know that in our network of workshops, some, are more efficient than others, and our aim is to continue to bring the efficiency of those at the lower end of the spectrum closer to the average. Additionally, to make continuous improvement in those which are best in class. It is therefore a process of continuous improvement for all of them, as the average is always being raised.

Each workshop has a budget, and responsibility for this is devolved to the Workshop Manager, he is responsible for delivering his customers requirements within budget, or better. A budget however, is only as good as it has been set, and is not necessarily a measure of financial productivity, beating a slack budget is not stretching, nor is it desirable. We need a way to set a realistic budget, but, also to be able to compare how one workshop's expenditure on his fleet compares with the next, or the average.

We have therefore developed a set of productivity and performance indicators, which are used to benchmark each workshop and include targets to achieve.
(see APPENDIX 1 & 2).

One of the first measures we have is staff cost by equated vehicle.

By using this method it is possible to compare the staff cost of a workshop with a higher percentage of small vehicles, to one with a higher percentage of larger vehicles. It is also possible to evaluate the effective use of overtime, as any higher cost of labour will be reflected in the overall cost per EVV.

Input for staff costs is taken from our payroll system and this is then used to produce full year effected staff costs, against a cumulative average Equated vehicle value. Cumulative average is used as the vehicle population varies during the year.

Another valuable measure is Non Staff cost per equated vehicle. Here we use a selected number of expenditure lines, those which can be directly affected by good management, from the workshop budget information supplied from our management information system to produce a Non Staff cost per EVV. Some of these items are: Parts and Accessories, Tyres, Contract Maintenance (accident repairs, component repairs, cost of recovery) etc. Again, cumulative averages are used.

Earlier in the paper I referred to equated vehicles per man and our planning assumption of 50. A key measure is Hours per Equated Vehicle Value. This is our main productivity measure, and can split into Direct and Indirect hours, this is to ensure that there is no 'dumping' of hours into, say; training, computer input, stores management, etc. Again, this is expressed as a cumulative average figure.

For example, two workshops could be operating close to our planning assumption of 50, say, 45 EVV/man. Each could, however, be using a different number of hours per EVV, for example, workshop 1, is using 33 hours per EVV, whilst the other is less productive at 40 hours/EVV. We would also need to check the relationship between Direct and Indirect hours, but this is our fine tuning to the EVV which allows us to measure, quite specifically, relative productivity of workshops.

I have also already referred to budgets. These are able to be, and are, related to Total Staff and Total Non Staff expenditure per Equated Vehicle value. This is useful in the budget setting process, to ensure that for a given workshop adequate funding is provided, whilst avoiding over funding being introduced to the process. We are also able to measure year on year achievements even if the fleet is increased or decreased. For significant fleet increase or decrease, we can make allowances for budget either to expect under or overspend.

Productivity packs are produced centrally and are distributed to the relevant managerial staff within the business on a monthly basis. However, not only do the Workshops see their own performance, they also see their performance on a National basis in the form of league tables.

(see APPENDIX 2).

A condensed version of the tables are displayed within each workshop, through this front line involvement much interest is generated in how productivity can be raised. The tables are displayed for information purposes, and each workshop manager ensures that they are used in this fashion through his team briefing sessions. They are intended to ensure everyone is involved, after all, the productivity of a workshop not only relies on the effective management of staff, but also by including people in sharing information and listening to their suggestions for improvement. Technicians are the biggest single factor in improving workshop productivity, they turn the spanners!!!
(see APPENDIX 1).

Until very recently, we had a Workshop Managers bonus scheme which was directly related to achievement of targets set around EVV's. This year it has been changed to reflect a number of other Business initiatives. However, it is worth briefly outlining the previous scheme which was set within the framework I have outlined.

Workshop Managers Bonus Scheme.

Percentage points of salary, up to a maximum of 12% can be awarded for achievement of targetry. This is split into four parts each of which offers a maximum of 3%.

Budget Element:

- Performance against workshop budget

    1% if the Budget is met.
    2% if the Budget target is underspent by 2.5% or more.
    3% if the Budget target is underspent by 5% or more.

- Performance against target indicators

    i) Direct time and indirect time - Hours per EVV.

    1.5% if this indicator is below O O hours per EVV.
    1.0% if it is between OO and OO/EVV.
    0.5% if it is between OO and OO/EVV

    ii) Total maintenance spend per EVV.

    1.5% if the indicator is below £O /EVV.
    1.0% if the indicator is below £O/EVV.
    0.5% if the indicator is below £O/EVV.

Quality Element:

- Meeting measured quality targets

Up to 3.0% of salary, on a scaled average score being achieved, based on the results of a Customer Satisfaction Survey as it relates to the individual workshops customers.

- Meeting individual quality objectives

    Up to 3.0% of salary based on achievement of 3 quality targets set by the Field Engineer responsible for the workshop.

The workshop managers bonus scheme was set up as being self funding from the outset, and is a major factor in maintaining quality and reducing cost through increased productivity.

However, the targets measured allow management information to be used to identify areas for further opportunity, and to sustain continuous improvement.

I trust that this has been useful in understanding how Parcelforce Worldwide have met the challenge of raising workshop productivity and performance; through continuous measurement and improvement, containing vehicle maintenance expenditure, whilst ensuring we operate well within the law, sustaining high vehicle maintenance standards and the essential levels of customer service demanded by our operation.

There are any number of methods used to measure and improve productivity and produce cost reduction, this is one example we know has been successful within Parcelforce Worldwide. I hope that this will encourage others to develop schemes tailored specifically to meet the needs of their own Business requirements, and to achieve their goals in an ever demanding and increasingly competitive market.

PARCEL FORCE WORLDWIDE

MTW PRODUCTIVITY AND PERFORMANCE

ANYTOWN MTW

ACCOUNTING PERIOD 03 / 98

ISSUE	ACTUAL	TARGET
STAFF IN POST	10	12
E.V.V. MAINTAINED PER MAN	62.60	50.00
FLEET AVAILABILITY TO OPERATOR	96.53%	98.50 %
SICK LEAVE	0.00%	< 4 %

MTW MAINTENANCE HOURS PER E.V.V
FULL YEAR EFFECTED

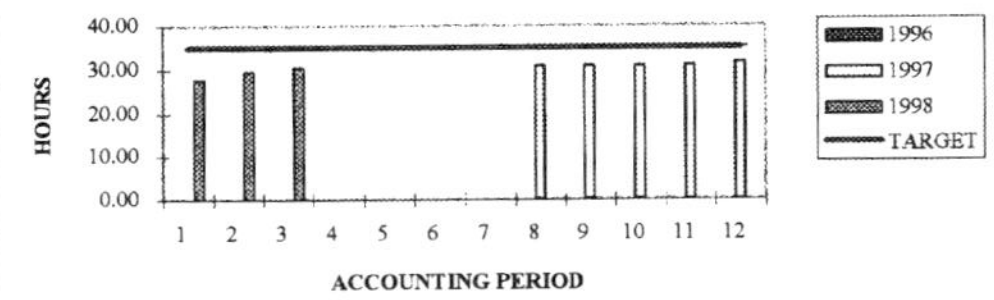

MTW STAFF COSTS - FULL YEAR EFFECTED

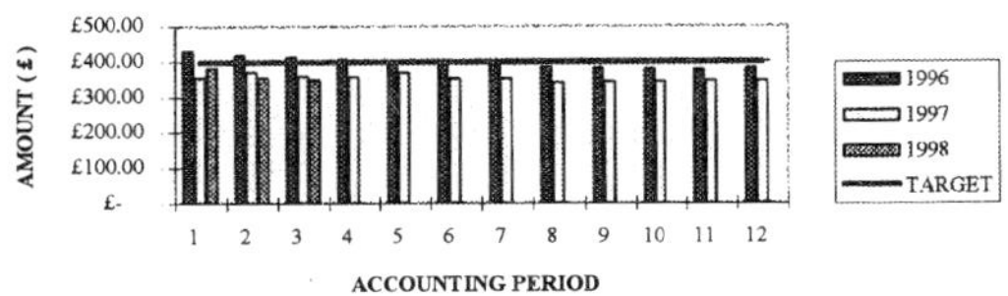

MTW EQUATED VEHICLES MAINTAINED
PER MAN EQUIVALENT

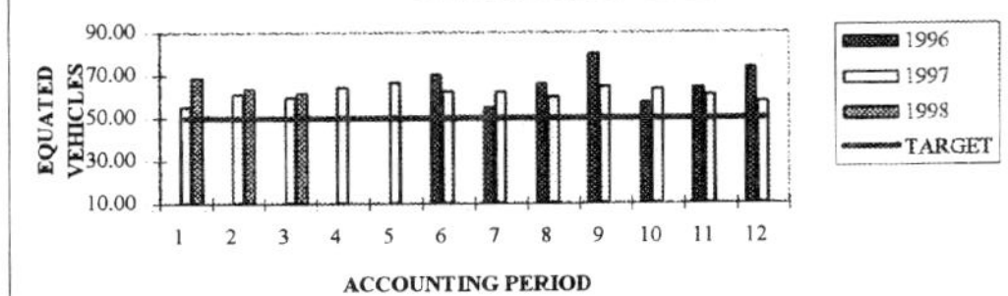

MTW NON STAFF COSTS - FULL YEAR EFFECTED

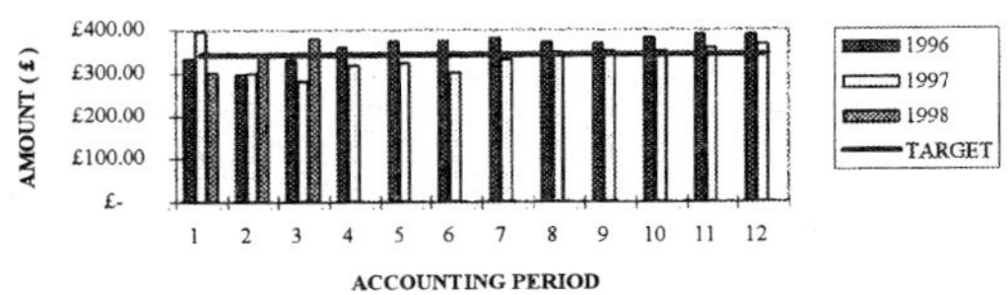

MTW OVERTIME AS % OF ATTENDANCE HOURS

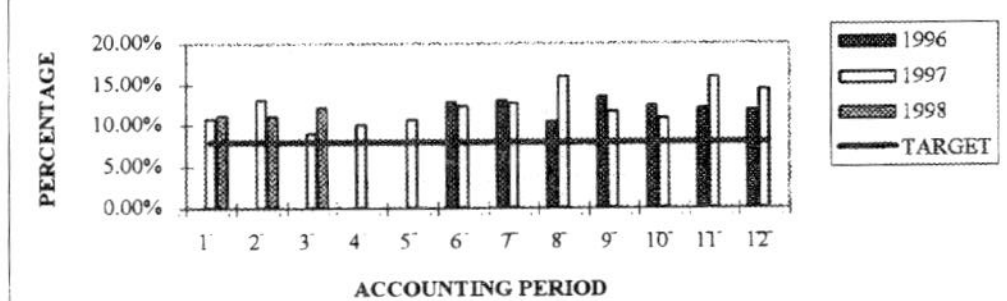

APPENDIX 1.

MOTOR TRANSPORT WORKSHOP PERFORMANCE SUMMARY
ACCOUNTING PERIOD : 3

TIME PRODUCTIVITY

MT WORKSHOP	MTWM	MTWP	UREQ	UNACCOUNTED HOURS	WAITING TIME	SUPERVISION AND ADMIN	STORES AND SHOPPING	COMPUTER TIME	PRODUCTIVITY DIRECT HOURS	FOREIGN WORK BALANCE	TOTAL	EQUATED FLEET	HOURS PER E.V.V. FULL YR EFFECT
BIRMINGHAM													#DIV/0!
BRISTOL													#DIV/0!
EAST LONDON													#DIV/0!
EXETER													#DIV/0!
GLASGOW													#DIV/0!
LEEDS													#DIV/0!
LIVERPOOL													#DIV/0!
LOUGHBOROUGH													#DIV/0!
MANCHESTER													#DIV/0!
MILTON KEYNES													#DIV/0!
PETERBOROUGH													#DIV/0!
REDHILL													#DIV/0!
SOUTHAMPTON													#DIV/0!
WASHINGTON													#DIV/0!
WEST LONDON													#DIV/0!
WINNERSH													#DIV/0!

TERRITORY	MTWM	MTWP	UREQ	UNACCOUNTED HOURS	WAITING TIME	SUPERVISION AND ADMIN	STORES AND SHOPPING	COMPUTER TIME	PRODUCTIVITY DIRECT HOURS	FOREIGN WORK BALANCE	TOTAL	EQUATED FLEET	HOURS PER E.V.V. FULL YR EFFECT
NORTH	0.00	0.00	0.00	0.00	0.00	0.00	0.00	0.00	0.00	0.00	0.00	0.00	#DIV/0!
EAST	0.00	0.00	0.00	0.00	0.00	0.00	0.00	0.00	0.00	0.00	0.00	0.00	#DIV/0!
WEST	0.00	0.00	0.00	0.00	0.00	0.00	0.00	0.00	0.00	0.00	0.00	0.00	#DIV/0!
BUSINESS	0.00	0.00	0.00	0.00	0.00	0.00	0.00	0.00	0.00	0.00	0.00	0.00	#DIV/0!

MTW FINANCIAL PERFORMANCE

STAFF COSTS

MT WORKSHOP	1997 / 98 COST PER E.V.V.	1998 / 99 COST PER E.V.V.	PERCENTAGE VARIATION ON PREVIOUS YR
BIRMINGHAM			#DIV/0!
BRISTOL			#DIV/0!
EAST LONDON			#DIV/0!
EXETER			#DIV/0!
GLASGOW			#DIV/0!
LEEDS			#DIV/0!
LIVERPOOL			#DIV/0!
LOUGHBOROUGH			#DIV/0!
MANCHESTER			#DIV/0!
MILTON KEYNES			#DIV/0!
PETERBOROUGH			#DIV/0!
REDHILL			#DIV/0!
SOUTHAMPTON			#DIV/0!
WASHINGTON			#DIV/0!
WEST LONDON			#DIV/0!
WINNERSH			#DIV/0!

TERRITORY	1997 / 98 COST PER E.V.V.	1998 / 99 COST PER E.V.V.	PERFORMANCE INDICATOR
NORTH			#DIV/0!
EAST			#DIV/0!
WEST			#DIV/0!
BUSINESS			#DIV/0!

NON-STAFF COSTS

MT WORKSHOP	1997 / 98 COST PER E.V.V.	1998 / 99 COST PER E.V.V.	PERCENTAGE VARIATION ON PREVIOUS YR
BIRMINGHAM			#DIV/0!
BRISTOL			#DIV/0!
EAST LONDON			#DIV/0!
EXETER			#DIV/0!
GLASGOW			#DIV/0!
LEEDS			#DIV/0!
LIVERPOOL			#DIV/0!
LOUGHBOROUGH			#DIV/0!
MANCHESTER			#DIV/0!
MILTON KEYNES			#DIV/0!
PETERBOROUGH			#DIV/0!
REDHILL			#DIV/0!
SOUTHAMPTON			#DIV/0!
WASHINGTON			#DIV/0!
WEST LONDON			#DIV/0!
WINNERSH			#DIV/0!

TERRITORY	1997 / 98 COST PER E.V.V.	1998 / 99 COST PER E.V.V.	PERFORMANCE INDICATOR
NORTH			#DIV/0!
EAST			#DIV/0!
WEST			#DIV/0!
BUSINESS			#DIV/0!

TIME PRODUCTIVITY

MT WORKSHOP	1997 / 98 HOURS PER E.V.V.	1998 / 99 HOURS PER E.V.V.	PERCENTAGE VARIATION ON PREVIOUS YR
BIRMINGHAM			#DIV/0!
BRISTOL			#DIV/0!
EAST LONDON			#DIV/0!
EXETER			#DIV/0!
GLASGOW			#DIV/0!
LEEDS			#DIV/0!
LIVERPOOL			#DIV/0!
LOUGHBOROUGH			#DIV/0!
MANCHESTER			#DIV/0!
MILTON KEYNES			#DIV/0!
PETERBOROUGH			#DIV/0!
REDHILL			#DIV/0!
SOUTHAMPTON			#DIV/0!
WASHINGTON			#DIV/0!
WEST LONDON			#DIV/0!
WINNERSH			#DIV/0!

TERRITORY	1997 / 98 HOURS PER E.V.V.	1998 / 99 HOURS PER E.V.V.	PERFORMANCE INDICATOR
NORTH			#DIV/0!
EAST			#DIV/0!
WEST			#DIV/0!
BUSINESS			#DIV/0!

APPENDIX 2.

S626/007/98

Parts operations from an engineer's perspective

S DAVIS
TNT UK Limited, Atherstone, UK

The Road Transport Industry has undergone significant changes in recent years and is still in a stage of continuous development. These major changes have been brought about because of the complexity of the modern vehicle and the requirement by the operator to maximise vehicle utilisation, which in turn reduces whole life operating costs. With specific regard to Commercial vehicles, increased legalisation has also been a contributory factor. The vehicle manufacturer has also influenced these changes by the mechanisms they use to sell their products into the market. In turn this has instigated a rapid transformation in the way that the modern fleet workshop hence Engineer approaches and meets the demands of vehicle maintenance.

In order for a modern vehicle repair workshop to cope with these changing demands and to deliver efficient and cost effective maintenance it is initially important to address 3 main areas:-

- Effective proactive management
- Suitably qualified human resource
- Supply of materials

In the absence of any of these 3 factors efficient maintenance will be difficult.

The purpose of this paper is to concentrate on the involvement of the Engineer in ensuring that the purchase and quality of components supplied are maintained to facilitate cost effective and efficient maintenance, whilst achieving optimum levels of vehicle availability.

The provision of spare parts into a workshop is considered by many Engineers to be a mundane task. In a number of incidences this is due to the lack of knowledge and understanding of the processes involved, coupled with the fact that additional administrative control measures have to be in place to guarantee the process is successful.

Apart from large facilities, storeman within the fleet workshop have largely become extinct. As a result of this the Engineer has had to take on the additional responsibility of the procurement and issue of parts in the workshop environment. When you take into account the diversity of parts required to feed a busy workshop ranging from bulbs to engines, it is an important area which needs to be approached in a systematic and competent manner.

Whilst in many large organisations, nationally arranged contracts have assisted in the commercial negotiation, the Engineering Manager still has a large role to play in the parts acquisition process.

Fleet profiles have changed significantly in recent years with many fleets now having vehicles from several different manufactures. This has had a dramatic effect on the diversity of components required to support vehicle maintenance. However, it is also important to remember that modern vehicles are more reliable and components are more durable (a point many operators do not necessarily agree with). This in turn causes a dilemma for the Engineer as he has to be in a position to anticipate component failures if he is to succeed in ensuring that effective maintenance is undertaken and that vehicle availability is maximised.

With specific regard to this area, the vehicle manufacturers have a significant part to play as they are in a much better position than the workshop Engineer to anticipate component failures. During the design and development process an assessment of the durability of components takes place and the results analysed to ensure that items will withstand the rigours of a vehicles operating environment. In recent years this product knowledge has enabled an increasing number of manufacturers to offer extended warranties (whether time or distance). The warranty package can only be arrived at by the relevant manufacturer fully understanding the durability of their products and components before they are in a position to add this benefit to their product marketing.

Whilst many large organisations have systems and processes in place to monitor recurring failures on vehicles, the smaller operator is not in the same position. However, effective stock control will assist every organisation to monitor and improve this area.

When considering the provision and control of stock we must initially take into account the physical area available to hold the components required. In many workshops space is at a premium, however, enough room must be made available to guarantee that components are held in a manner to ensure the parts are retained in good condition and to allow for easy identification and retrieval. This factor can be overlooked, but it is an important issue to be aware of.

Other factors which influence the provision of a suitable area for stock i.e. lighting, heating, insurance, security and racking must not be neglected.

Another point to consider is stock obsolescence as this must be brought into the equation when overall costs are taken into account, however, careful control and monitoring should minimise the impact of this issue.

It is important to have an understanding of all the reasons why stock holding is necessary in the workshop environment and why it contributes to efficient workshop management, which in turn assists with maintaining high levels of vehicle availability.

In most Commercial vehicle fleet workshops working area is at a premium, consequently the efficient use of the available floor space is essential. Moving disabled vehicles in and out of the workshop is both time consuming and inefficient. Whilst in all cases the requirement to undertake this is not always down to parts unavailability, it is in many incidences a contributory factor. In fact, with large vehicles the reassembling of failed components is sometimes required to make this possible. This becomes an even more important factor when the workshop is operating on a 24 hour cycle, as the different shift patterns may require specific work to be undertaken which will require the workshop to be cleared before each shift commences.

For the individual mechanic continuity and motivation is also an important consideration, as a mechanic does not want to start and stop repairs whilst awaiting parts. This becomes more frustrating if the part is relatively small and does not require a great deal of time to rectify. Whilst these areas are not exhaustive it does demonstrate why parts supply, hence stock holding is an important issue.

During recent years the Commercial vehicle dealer network has shrunk, therefore, the remaining dealers now have to cover a larger geographical area. Depending on the location of the independent workshop this has resulted in increased delivery times and made distress purchases more difficult. Whilst a proportion of manufacturer dealer networks have introduced secondary agents to alleviate this problem, generally this has not counteracted the overall reduction in the number of dealers hence stocking points.

Another area which has impacted on stock is the "just in time" concept, where manufacturers distribution warehouses have been closed or have been consolidated into larger central locations. This has then resulted in components being fed to the stocking points on demand. As you will appreciate this has meant that the requirement to forward plan and anticipate part usage has increased throughout the whole supply chain. The link between the workshop, stores and the vehicle is the last in the chain and is the most critical for the Engineer.

What is stock ?

Having considered the reasons why the workshop requires stock, it is important to understand what exactly constitutes stock.

The definition of stock is:- all materials required by a workshop to support either the workshop, or the vehicles maintained. This will include lubricants, consumables, tyres and spare parts. Whilst these areas are specific in stating what constitutes stock within a vehicle workshop environment this can be applied to any stocking process.

What are the reasons to hold stock ?
The reasons to hold stock items are various and will differ at each maintenance location, however they will include:-

- Vehicle parc
- Vehicle duty cycles
- Operator requirements
- Opening times of workshop
- Local supplier support

Once this is fully understood we can then move forward to obtain a better understanding of why the Engineer requires stock and its effect on workshop performance and efficiency.

Why does the Engineer need stock ?

- The convenience of having things available as and when required without making special arrangements
- Cost reduction through purchase of optimum quantities
- Protection against the effects of forecast error
- Inaccurate records
- Mistakes in workshop planning
- Provision for fluctuations in unscheduled repairs
- Components which are difficult to obtain

To summarise, the principle purpose for workshop stock holding is to create a buffer between supply and demand. The systems detailed in the following pages are to ensure that the materials are available in the right place, at the right time, in the right quantities.

Once the reasons for holding stock are understood, the next aspect to take into account is stocking policies. This is a major factor and must be fully grasped as the composition of components held is paramount if successful and efficient maintenance is going to be achieved.

A major influence on stocking policies is outside the direct control of the workshop, this being the supplier network and in particular the suppliers in the immediate vicinity. These suppliers will be made up of manufactures, franchised dealers, nationally negotiated contracts (where applicable) and the factor network.

From an Engineers perspective there is a dilemma as he is faced with two differing scenarios; the first one is to be a typical Engineer and be rather safe than sorry which results in excessive stocking policies, the second is that the more stock held, the more there is to monitor and control.

In an obscure way this can often be decided by financial constraints as every Engineer is tasked with reducing the stock holding to an absolute minimum without affecting workshop efficiency which will ultimately have an impact on vehicle availability.

In general, stocking policies will be divided into 3 main categories, these being:-

- Fast moving service parts

- Components prone to abuse or failure
- Components purchased specifically for vehicles which are to be fitted at a later date (before next scheduled maintenance period)

This is further complicated when the various makes, models and ancillary equipment are taken into account. In many incidences even across differing manufacturers, components can be cross referenced which will have an affect on the composition and quantity of stock held. With careful monitoring and control it is possible to achieve good stock keeping without sacrificing workshop efficiency.

With reference to the financial controls, impressed or consignment stock can be a viable option, but it must be remembered that in most cases the ultimate responsibility for the stock held within the workshop falls under the control of the Engineering Manager.

When taking all these factors into account the whole issue of a workshop parts operation can appear to be complex, but as long as basic rules are followed the efficiency of a workshop and its support of vehicle maintenance can be significantly improved.

It is of course imperative that a solid platform is in place to control the acquisition and flow of components through the workshop. It is also necessary to ensure that this system has the ability to allocate components to vehicles, which in turn assists with accurate vehicle maintenance costing. The allocation of parts from a cost point of view will be different according to the requirements of each individual company, however, it is important that this flow of components through the workshop is monitored.

This platform is in the form of a stock control system which can be manual or computer based. This system need not be of a complex design as the quantity of individual items held, coupled with the stocking policy will dictate the sophistication.

Manual stock systems are still popular and perform adequately, however, electronic systems are on the increase as they allow an interface with other workshop systems which reduces the duplication of time and effort. Additionally, the electronic system will give improved reporting capabilities with regard to usage and re-ordering.

The manual systems, can be simply a duplicate book, a T card system or an Industry recognised paper system. For the smaller workshop these manual systems have provided a good level of control for many years and there is no reason why they should not continue to do so.

Computer based systems can be PC based for single sites or fully integrated packages which allow various sites to enquire on both stock levels and descriptions, and where necessary transfer stock items. Until recently this benefit has only been available to the vehicle manufacturer. It is also possible to produce baskets of parts, this then removes the need to book out individual components when undertaking a repair activity which necessitates the use of a consistent number of parts. A typical example of this is when a scheduled maintenance is carried out as all the parts can be charged at once. Whichever system is chosen it is imperative that it is robust to capture all facets of stock control.

To further understand the fundaments of a stock control system it is important to be fully aware of the make up of this system. With this in mind it is prudent that time is spent detailing the main requirements for this system to accommodate.

Firstly, all parts have to be clearly identified, this is achieved by the unique part number or a description which can includes the type of vehicle. The latter is more popular as different suppliers can generate their own numbers for the same part. Whichever method is chosen it must be consistent across the approved system to ensure that continuity is maintained.

It is also necessary to segregate parts within the stock room, this is normally achieved by a bin location. This can be physically a container to retain the item or for larger parts an area within the stock location. In the main this is the method used by the mechanics to allocate parts to a vehicle or repair job.

Maximum and minimum stock levels are a major consideration when setting up a system as this allows controls on the amount of stock held against the level for re-order. Depending on usage, these quantities will be amended on an ongoing basis.

The usage figures are of paramount importance to ensure that stock levels are maintained at the required levels, and it is possible to identify that repetitive repairs are being undertaken which may affect warranty claims which ultimately will have an effect on maintenance costs and vehicle availability .

To enable the flow of materials to be monitored through a workshop it is necessary to control the component throughput. This is achieved by accurately recording the parts in and out of stock at the bin location level. The utilisation of this method enables a numerical check to be kept on individual component usage.

Both the purchase and selling dates should also be recorded within the system as they are necessary for audit and where applicable warranty identification. Many suppliers accept that the warranty clock will commence on fitment and not purchase as long as the necessary proof can be provided.

It is also advisable to include the supplier within the system as this will aid the assessment of the lead time and assist with audit and administration. If different suppliers are used for the same component it is important that the system can distinguish between them.

Carrying stock for a workshop is expensive and as previously stated an Engineer will tend to overstock to cover all eventualities. As a result, it is necessary to constantly strive to reduce stock in order to combat the natural tendency to increase. Constructive approaches to stock reduction include:-

- Arranging for things to be delivered just in time instead of stockpiling just in case the need arises
- Devising ways to reduce stock ordering, set-up costs and lead times so that optimum quantities are smaller
- Making forecasts more accurate, ensuring that records are correct, ie better planning

One of the main disadvantages with these approaches is a general increase in paperwork which invariably leads to an increase in administration, however, it should be remembered that with careful controls this can be managed at an acceptable level.

Another increasingly popular way of providing stock into the workshop environment is impressed or consignment stock. As this type of stock is not reported on the balance sheet it allows the Engineer to maintain higher levels. Other benefits of this alternative would be a reduced number of suppliers to the workshop, which in turn reduces the administration time required. There are however some disadvantages with this option. The mechanics are responsible for removing their own parts from the stores, this can lead to diagnosis by substitution which can prove inefficient and expensive. Another point to remember with this option is that you have in the main committed to a single source and sometimes this may not always be cost effective as you are invariably tied into this supplier. In turn this may also reduce the relationship you have with other component providers in the immediate vicinity, which is an important factor to take into account when their is a requirement to make a distress purchase outside of the consignment stock supplier.

The role of the component supplier should not be ignored and good communication maintained with them as inevitably the impress or consignment stock route will not meet all the needs of the workshop.

As previously stated modern fleets are very diverse, consequently the number of suppliers has also increased which in turn increases the administrative burden on the workshop. A typical workshop in TNT will have approximately 80 suppliers which includes a secondary source. These suppliers will be made up of nationally negotiated contracts, manufactures, franchised dealers and local factors. All of these suppliers will contribute to the smooth operation of the workshop in support of cost affective and efficient vehicle maintenance.

To finish, this document demonstrates that the control and flow of components obtained by any source through a workshop is paramount to achieve effective maintenance. Additionally, the system in place need not be overly complex in its design or construction to achieve the required objectives. The important factor to consider is that the chosen system must be monitored to ensure that workshop efficiency is maintained which ultimately results in a high level of vehicle availability. As with other processes compromises will need to be made to tailor the process to individual workshop and operator requirements. It is also imperative that all staff working within the workshop are fully aware of and understand the operation of stock control.

S626/008/98

Computerized systems

S OATES
Oakrange Engineering Limited, Rotherham, UK

OBJECTIVE

To assist in developing vehicle repair and maintenance workshop operations with minimum vehicle downtime, maximum repair reliability, at optimum cost.

INTRODUCTION

Martin Croucher – Sector Manager for Intermec technologies.
Responsible for the Logistics sector within Intermec, Martin has some 15 years experience of implementing mobile data capture solutions within the fields of manufacturing, retail and distribution. His main focus is currently providing solutions within the supply chain involving the implementation of radio data terminals within warehousing and on-line truck fleet management in the road logistics sector.

Steve Oates – Managing Director OE technologies.
Steve is the Managing Director of OE Technologies who have been working in the field of data capture, communication, analysis for military and commercial clients for more than a decade.

BRIEF OVERVIEW

Intermec are a global player in the data capture market, they are part of Unova Corporation of the USA, they have a turnover of around $1.5 bn. More significantly in today's context, they have over ¼ million mobile engineers hand held and in -cab data terminals installed throughout the world. Martin is heading up the UK operation to develop these technologies into the logistics sector.

OE technologies are essentially an R & D organisation, involved primarily at the 'far side' of the supply chain, providing technology to the vehicle manufacturers, dealerships and operators. Part of these activities is involved in the black art of calibration for workshops, tachograph stations, etc. Another area is in the development of in-cab data-capture technologies, hence the tie with Intermec.

Just before we launch into the practicalities of utilising computers in the area of engineering management, there are a few elements of jargon to consider; – We will generally talk about 'data capture'. This implies the use of an electronic device at the 'coalface' i.e. where the work is taking place. This may be communicated in 'real-time' (i.e. as it happens) via a wide area network (wan), such as the GSM mobile data telephone network, or, downloaded into a PC on a 'batch' i.e. per job, or, shift basis.

The data may be a manual input e.g. a button press, or, more common, a touch screen, or, a bar code, or, in the case of a vehicle, an automatic input from the vehicle instrumentation or engine management system.

That 'data' will normally be utilised by an 'application' on the PC to turn the 'data' into 'information'. For example; the technician touches the 'end of job button' on the data capture terminal. The data is passed to the PC, which updates the job record to confirm it is finished, the application may then automatically compute the time and cost against that job.

The reciprocal of this process is that the host PC can then pass the details of the next job back to the mobile terminal for the technician to carry out. This may include specific instructions, parts lists, standard times etc.

As a simplistic statement, any proprietary workshop application operating under 'windows', should be able to accept data from a mobile terminal via a simple integration program that links the two components together.

The general concept is that you are feeding information into a relational database that enables you to record, retrieve and analyse the information typically by date, truck, technician, etc.

THE PRINCIPLES OF DATA CAPTURE INCLUDE:

Distributed Data Input

Technology offers the ability to allow workers to update and interact with central computer systems at the 'coalface'. This is due to the high ruggedness levels of data input terminals, coupled with processing power and the wide range of interfaces to central systems. Data input terminals may be used within site premises, on-line to the computer system via fixed or mobile networks, or, may be operated geographically on a wide area network.

Empowerment

Placing computer aided applications at the front end of the business, allows employees to complete their duties with maximum efficiency and minimum reliance on the availability of management for day to day information. Customer and company data is available 'in the hand' to assist decision making processes and maximise business. For example, a workshop engineer can see at his work station, product maintenance history, parts availability, diary resources. His activities are recorded directly onto the central computer, system for colleagues, with reference to re-ordering, resource allocation etc.

Speed

Reporting work activities, as they happen, directly onto the central system allows realtime updating and eliminates the delays of traditional paperwork systems, where the company is reacting to information which is hours or days old. The additional use of machine readable technologies such as barcodes, means that data can be entered on the press of a scanner button, rather than laborious keying and allow more time for revenue based activities, such as vehicle repair.

Accuracy

Information keyed into a system attracts errors due to keyboarding skills and the misreading of colleagues handwriting. For example a skilled keyboard operator is expected to mis-key 1 in 300 characters. When dealing with 10 digit product codes, this relates to 1 in 30 incorrect SKU's. Barcode scanning can reduce these errors to less than 1 in 3,000,000 errors.

Administration

A whole layer of expensive manual repetition is potentially reduced by having the person responsible for a business activity directly update the systems, rather than have operations based on collecting work sheets and keying in information as a later activity.

THE APPLICATION OF AUTO ID TECHNOLOGIES

Barcoding

- Lowest cost of auto ID, (0. 1p to £1.00) conventional and specialist printing.
- Paper, label fabric and metal materials.
- 'Write once, read many times'.
- Limited number structure (up to 30 characters).
- Keyword to a database lookup table. Line of sight scanning.
- 2D barcodes allow 100's of characters, increase resilience and allow the database to be within the barcode where access to the computer system is unavailable.

RFID

- Higher cost technology (50p to £100), pre manufactured.
- Electronic tag, packaged in materials suited to cost/hazardous environment requirements.
- Read and write technologies available.
- Can contain 1000's of characters.
- Does not require line of sight, can be embedded in 'non-metallic substrates.

AN EXAMPLE OF THE USE OF 'ON-LINE' DATA CAPTURE WITHIN THE TRANSPORT SECTOR

Radio data terminals, on a mobile local area network in the warehouse, control and report on picking and put-away activities. Once a product or service is loaded onto a vehicle, mobile tracking of data helps increase driver and destination dock productivity.

Information transferred during the day updates all aspects of the operation:

- Proof of delivery/invoicing.
- Delivery window accuracy.
- Vehicle maintenance requirements.
- Asset utilisation information.
- Operating costs.
- Asset tracking (roll-cages/containers).

Live examples of the application of technology, together with proven benefits best demonstrate the successful use of data capture:

∇ **URS Logistics** (formerly United Refrigerated Services)

> Largest U.S. distributor of refrigerated goods
> Continuous tracking of pallets from receiving
> Radio Computers on forklifts
> Real-time tracking, under sustained harsh environment
> Improved efficiency, productivity and flexibility
> Decreased labor needs/costs - one shift eliminated

∇ **Euro-fleet**

> Distribution of over 1000 cars/day.
> 40,000 car stock holding.
> Radio data terminals on three sites.
> Stocktaking reduced from 60 man days/month to 8 man days.
> Faster vehicle processing time.
> Improved accuracy.

∇ **Pepsi**

> Wireless data solution for order entry and delivery
> Hand held computers
> 24-hour order turn-around
> Two-hour order transmission improvement

∇ **Ciba Speciality Chemicals**

> Lot traceability and billing control for bulk chemicals
> Rail-car data entry via networked terminal
> Improved accuracy, reliability
> Reduced material handling time

∇ **Chrysler**

> Inventory tracking system
> Radio frequency network integrated into Chrysler computer system
> Reduced time and cost spent on inventory processing
> Increased accuracy in warehousing

∇ **Black & Decker**

> Producer of lawn mowers and power tools needed to improve receiving and inventory control process
> Radio data terminals and printers
> ROI - seven months
> Increased inventory turns, reduced labor cost, improved accuracy

∇ **U.S. Department of Defense**

> Total asset visibility
> $250 million AIT contract
> Use of auto ID and portable terminals
> Operational efficiency
> Tactical mobility
> Global responsiveness

∇ **Kwik Fit**

- 149 vans for mobile tyre service fleet
- Hand held computers for technicians
- Reduced paperwork
- Operational efficiencies

∇ **Tibbett & Britten**

- 3500 vehicles in UK
- Sainsbury's with Chronicle fleet management
- Warehousing with RF terminals Increased customer service
- Reduced operating costs
- On-line management

∇ **NYNEX**

- Installation and maintenance of regional telephone system
- PEN*KEY mobile computers for retrieval of information by field technicians, and transfer data via modem to switch stations
- Field technician productivity increased of 1000%
- Labor resources maximized
- Improved network efficiency
- Fewer service interruptions

∇ **Securicor Cash Services**

- 60% of UK Cash-in-Transit.
- 30 million containers/annum.
- Radio data terminals in 62 depot vaults.
- Real time tracking accuracy.
- Company wide information increases customer service efficiency, proactive, not reactive.
- Capability for 50% increase in throughput.
- Faster, error free cash handling.

∇ Summary

Choose the right components for the application:

- Data entry terminal:
  - Fixed station.
  - Portable batch device.
  - Mobile on-line terminal.
- Application software.

- Network:
  - Fixed/mobile local area network.
  - Fixed/mobile wide area network.
- Automatic identification technology:
  - Barcodes.
  - Radio transponders.
  - Smart cards.
  - Magnetic stripe.

Obviously trying to cover an entire industry in such a brief period can only give you a flavour of just what is available.

The next part of the presentation is to give you a 'commercial appraisal' of the applications of the technologies, and then to take things just a little further to show just what is happening at the leading edge

First, let us be pragmatic, just what is cost effective for a contract hire company spending £10 m per year on vehicle maintenance, is not the same as a tightly run own account fleet of 50 trucks. In the former case, it may well prove cost effective to spend £2 m to have a totally integrated system.

Just to give a real case, a colleague of ours operates ¼m cars throughout Europe, he has a customer call centre with several hundred staff essentially dedicated to 'trying' to obtain current 'accurate' mileages for maintenance planning purposes. (If you care to do the numbers, that is at least 250,000 calls per month, which is approximately 10,000 calls per working day !)

An integrated system whereby the vehicles automatically record and report in mileages over a radio network is a cost effective option to his business.

In the latter case, where there are 5 technicians very much 'hands on', it may prove to be more cost effective in having a stand alone P.C. application with perhaps a single data capture terminal for technicians on the shop floor, one for works van/breakdown use and a terminal in the stores for parts issue.

In the broader context, I will offer you a few statements that have very significant impact on our 'business' of fleet engineering;-

10 years ago, we ran a seminar for the great and the good, looking into the 'crystal ball' for future areas of impact of technology in fleet engineering. For me, the 'impact' was not about 'technology', it was, in a way, anti-technology!

In that seminar, a colleague of ours, the human resources director for a truck manufacturer, stood up and pointed out that the technological developments we were forecasting were unlikely to be taken up. Popular man!

What he pointed out, was that; as the demographic analysis of the population showed, there were to be 25–30% fewer school leavers in the forthcoming decade, as 'vehicle technician' was not high on the desireable occupation list - no apprenticeship, poor working conditions, etc. We would struggle to fill the human resource requirements. Any school leaver who was computer literate could command a job in an office rather than on the hard shoulder!

In the last year there have been a few further statements that really made me think, apologies if they are a little plagiarised.

The first is from Charlie Secker the long time engineer for Whitbread plc.
Charlie reported his financial director told him to 'come out of the workshop and into the business'. Charlie presented an excellent paper at BTAC on just what he had achieved in applying the intent behind the statement.

The next is a statement from Steve Kirk of Volvo truck ; he stated; in recent years Volvo has moved away from being a product-lead manufacturer to being a market lead supplier.' For those of us who have been involved with Volvo for 25 years, that thinking marked a quantum leap in our industry.

The final statement is from Lord McClaurin of Knebworth, the Chairman of Tesco, who indicated that Tesco spent £1.4 bn moving goods to their stores. A big number.

The seminar and those three statements have a strategic impact on fleet engineering. In particular, just where the technologies can be, and should be, applied.

As stated right at the beginning, my company are involved in R & D. We are looking in terms of just where we can make a commercial difference. After all, we do operate on a commercial basis and do not dream up these technologies just 'because we can'! What I will share with you is a little gem of empirical data, collected from hundreds of vehicles, operating predominantly on RDC distribution, on a single shifted basis.

It is, 70; 10; 20.

70% of the time i.e. In a 7 day week a vehicle is parked.
10% of the time it is parked with the engine running
20% of the time it is moving. (not all of that time is it loaded)

Now, compare that asset utilisation with any manufacturing process you wish to name; beer, yoghurt, cars, - any product will do.

The production engineer would get the sack at less than 90% Asset utilisation. So, we add all the overhead, labour, material, service and profit, to the product, then put it out on a vehicle with a productive asset utilisation of 20% (and still we cannot get the vehicles in for planned maintenance!)

Like I said this is going to be contentious. To give a little hope this situation is not the fault of the fleet engineer! An RDC is geographically 'where it is', because the drivers hours legislation determines a driver can only realistically journey maximum 4hrs 'out' and 4hrs 'back'. Common sense, and a decent routing package will site the RDC at the centre of something like a 2½ hr radius in relation to the stores.

2½ hrs 'out' and 2½ hrs 'back' is total 5 hrs moving.
20% of a day is 4.8 hrs. Q.e.d.
Loading, unloading, waiting (with the heater on) 2½ hrs, is 10% of the day.

The cynics amongst us may pass an observation that, that period of time just happens to make up the balance of the paid driver shift.

Just imagine:- an RDC, operating 100 trucks on a 50 hrs driver working week, doing 25 drops per vehicle, per week. Just 5 minutes 'won' per drop, would facilitate an extra drop per week. That extra drop would be a 4% improvement in asset utilisation. That would be 4 fewer trucks at a saving of say £¼ m per annum.

Compare that with what the hard pressed fleet engineer may be able to 'prune' from the maintenance costs of that operation and it becomes more evident that the savings are realisable in 'the business' not just the workshop.

The fleet engineering that 'wins' the 5 minutes per drop, may well be in the specification of the vehicle bodywork, it may be in the load docking equipment at the store, it may simply be an on board data communication device that allows the driver to forewarn the store of his e.t.a.

What we are trying to point out here, is that there is always a clear added value in being able to capture 'real' operational data quickly and accurately and then utilising that data to generate information to manage the business. This may be on the shop floor, the roadside repair site, the parts stores, the customer call centre or, from the actual vehicle.

The trick is to recognise just where the information is generated and investigate the application of the appropriate technology in the light of 'real information', on a cost effective appraisal basis. This may start with manual input into a simple relational database such as ms access, it may be a proprietary package for fleet management, it may be a bespoke piece of software, each one of these may then be fed with data manually, by devices or automatically according to the cost benefits identified.

In conclusion, there is more than enough technology out there, working in a variety of environments, that we can bring to bear on fleet engineering.

As to whether or not that is commercially viable, directly in the workshop, through data capture terminals, barcodes etc or, through remote vehicle diagnostics, or, through a broader perspective of the data capture being part of a strategic tool to identify cost effective applications, is quite definitely part of the fleet engineering aspect of improving asset utilisation. We forecast this will be the 'driver' for the next phase of technology application in our business.

If you really wish to look a little deeper into the crystal ball, I could identify the increasing reliability of vehicles (really!), the ecological issues of pollution caused by vehicles, the environmental impact of more vehicles with no more roads, the effects of education and demographic analysis of the workforce, also the effect of improving and lower cost of communications. All of which are going to impact the fleet engineering function at an accelerating rate. Fact.

In my career, I have consecutively seen; the age of the fleet engineer running the business, the operations running the business, the accountants running the business, and now the IT department running the business.

What goes around, comes around?

S626/009/98

Counting the costs

G SPOONER
Vehicle Maintenance, Metropolitan Police, London, UK

I recall many years ago engineering being defined as the art of compromise. As engineers, we learn to compromise early in our careers. Such as between strength and weight, function and maintainability, and cost and availability. You want a high performance vehicle? It will cost you either huge development and production charges to build a sophisticated piece of equipment to achieve high efficiency, or high fuel bills, or complex repair operations if it goes wrong. Maybe all of these factors!

The effect of cost was therefore a factor ingrained within my early career. But as a young engineer, I first thought that finance was just figures, not an art, let alone one of compromise. Whilst that is generally true of *financial* accountancy where the figures are what they are and have to be reconciled, I soon learnt that *management* accountancy is completely different.

Management accountancy, I contend, <u>is</u> an art and <u>is</u> definitely a compromise. This means that <u>you</u> as an operations manager whether you are an engineer or of another non-finance discipline, not accountants, can set the basis for how your organisation charges, and the rates for the work you do based on operational experience and knowledge. For often it is only those in the front line that understand the dynamics of the service to be provided, the cost of providing that service, and how best to maximise income.

This presentation attempts to break down the relevant factors into something that you can all understand. It is not about finance but about costs and charges and plain common sense.

Let me state five principles:

1. Monitoring & Control - To control something, it has to be monitored; therefore the monitoring of costs and the resultant performance is a fundamental requirement of management.
2. Profit - income must exceed costs, although over what period is another matter.
3. Cost recovery - there is no direct relationship between costs and charges.
4. Accountability - you cannot be held responsible for costs or charges which you cannot control.
5. Reconciliation - costs and charges must be reconciled across the whole organisation.

I will explore each of these principles during this presentation and endeavour to justify them. Most of the presentation will concentrate on monitoring through recording costs in a suitable way so that meaningful information can be reported. I will also briefly cover performance indicators and benchmarking although these could take a complete paper by themselves.

The examples I will use relate to vehicles and their maintenance but even within what would seem to be a fairly standard industry, there are no doubt many differences in the detailed approach to what I am presenting. However I believe that the general principles set out here are sound ones with which most will agree. It will be interesting to hear feedback from those involved.

MONITORING & CONTROL

Cost control is a fundamental principle of management. Without knowing what the costs are, it is difficult if not impossible to control them. If you can't measure something, you cannot control it. This is a very important message for you to remember because although it sounds, and is, simple, I have come across many cases where this basic message has not been appreciated. So costs must be monitored. How this is done can be described as an art.

All organisations have a financial accountancy system, normally (but not always) computerised, in which all costs and payments are recorded. There may also be other modules which link in to the general ledger sections of a computer package. Parts for instance may be controlled by a computerised stock control or inventory management module. But it is where costs are allocated that is important as this enables management reports to be generated. If the costs are allocated to what we can call *expense codes* in a meaningful, logical way then meaningful reports will be possible. And without meaningful reports being generated and used, the time spent monitoring costs will be a complete waste.

As an operating engineer, I incur costs in providing a service. The main items are:

- Staff
- Materials
- Plant and buildings

Staff costs can be separated into the following typical expense codes:

- ⇒ basic pay
- ⇒ overtime
- ⇒ allowances, such as:
    - ◊ location allowance
    - ◊ shift allowance
    - ◊ travel and subsistence expenses
- ⇒ ERNIC (Employers Related National Insurance Contributions)
- ⇒ Pensions
- ⇒ Staff support costs such as training

All staff costs are gross pay - these are all the organisation's costs of employing staff (but not the <u>employees</u> pensions and national insurance or any other contributions or deductions)

Other costs are:

- Parts
    - ⇒ stock - <u>are</u> these a cost? More of this later
    - ⇒ ordered for a specific job
    - ⇒ consumables, of which more later
- Special tools for repairs
- Technical information - manuals and microfiches etc. (and now CD-ROMs and internet links)
- Buildings and associated support services
    - ⇒ security
    - ⇒ cleaning
    - ⇒ maintenance
    - ⇒ insurance
    - ⇒ waste disposal
- Maintenance of workshop equipment
- Purchases of minor workshop equipment, say less than £1000
- Depreciation of capital costs - the change in value during the year of those items of plant or building of significant cost (say over £1000) which still have a significant tangible value at the end of each year, which would include:
    - ⇒ workshop equipment
    - ⇒ computer systems
    - ⇒ vehicle wash installations
    - ⇒ workshop support vehicles such as breakdown trucks
- Company overheads
    - ⇒ directors and their support costs
    - ⇒ sales and marketing
    - ⇒ research and development etc...

  or at least a contribution towards these by each cost centre within the organisation

There may be other costs which your organisation has identified as needing to be separated out and monitored. There are no hard and fast rules about how to structure these costs. For instance, <u>training</u> can be included as a staff cost or it could be kept totally separate.

Maintenance and support of computer systems may be costed direct to each business unit, or there may be a corporate system for the whole company to which each unit needs to "contribute". Often the company accountant or its auditors will recommend certain accounting structures so that similar costs are allocated to each part of an organisation, perhaps in relation to the staff numbers, staff costs, floor area or some other, hopefully equitable and logical, factor although rarely completely satisfactory to everyone.

Where do you allocate miscellaneous costs, i.e. costs that maybe only arise once in a blue moon or are small and do not fall easily into one of the other expense codes? I would ask for a miscellaneous code as otherwise it encourages costs to be "hidden" elsewhere, often in the biggest spending line.

Stock parts are paid for when you buy them but, until parts are used, they are an asset, a current asset (as opposed to a long term capital asset depreciated over more than one year). These have a value; either cost of purchase, average cost, or cost of sale, whichever your organisation defines. A stock control system, manual or computerised, will account for these parts and allow you to write off obsolete, damaged, or "lost" parts (sometimes known as "shrinkage"). All of these are costs to the business. How often these stock adjustments are made is a matter of judgement, or company policy. This can be done monthly to keep the system up to date perhaps through a perpetual audit, or annually as part of a complete stock audit. Either way there is a risk that parts that have been written off after stock checks then turn up again. Tools and workshop equipment need to be included in such checks or as a separate activity as any write- offs or replacements are all part of the workshop costs. Some of this will not apply if imprest stock is used which is essentially parts loaned by suppliers and paid for only when used although "shrinkage" can still occur and should be allowed for.

Control and charging of consumable parts is a sensitive area. These are parts which are of low unit cost or impracticable to charge separately but when added up amount to a fair sum which has to be recovered somehow.

Typical items range from abrasive paper through fuses and nuts to welding rods. Administratively it is impossible to account for each individual item if it has to be done manually. Even if a computer system is used, there is a practical limit to the number of parts that can be listed on an invoice. It not only looks wrong to the customer but also wastes paper. A management report should be run say, monthly, detailing those items which are not being costed direct. If consumables are not checked, the use and cost will creep up. As a starting point, the cost of consumables could be kept to anything less than, say, £5 unit value or, in total, less than 5% of the total parts costs. If the total looks high, look at the high cost items - maybe specialised bolts or bulbs - as these can be added to invoices without significantly increasing the invoice length compared to the cost recovery. Or look at the largest volume as this may be easier to reduce. This is an typical application of the Pirato rule where you look at a small percentage of what is creating the largest effect or cost. You will of course be keeping a check on "attractive" items to ensure that these parts are not being "misappropriated".

The costs that are left can be added as a standard charge per job to each invoice as a separate line at the end - i.e. *Consumables £3* - but this can often raise queries particularly when the job is one where the customer cannot see where consumables can be used. As a customer I don't like to see this on an invoice so I recommend that the overall cost is recovered in the hourly rate and kept as low as possible.

Returning to staff costs, we can also separate them into the following to give a ratio of direct to indirect staff costs:

- Productive staff - direct costs
- Indirect staff costs
  - ⇒ supervisors
  - ⇒ parts support
  - ⇒ drivers
  - ⇒ cleaners etc.

Textbooks suggest that all costs can be classified as either as fixed, semi-fixed, (or semi-variable) and variable. And so they are in a simplistic sense. But we all know that business is not simple! All I know is that I can control some costs short term and some costs I cannot even control long-term. Which ones are which, to which expense code they are allocated and how they are used is, I hope, common sense.

INCOME

All of these costs need to be recovered from the customer, whether that customer is an internal cost centre (or cost centres), or real paying customers from other organisations.

There are three basic options in charging for vehicle maintenance and repairs:

- Whatever the job costs in hours at an hourly rate, plus parts, and sub-contracted work. The cost can be based on the time it takes at a set hourly rate, or estimated using standard times if appropriate, but make allowances for non-standard equipment.
- Cost per vehicle per period (say, monthly over an agreed number of months as in a contract maintenance agreement)
- Cost per mile of vehicle operation

However there are risks if any one of a number of factors change, such as:

- What if the market collapses?
- Or you lose key members of staff?
- Or there is a technological change to the types of vehicle you are maintaining which significantly increases the service interval or parts requirements?
- What if you cannot react quickly enough to maintain an effective business?

Can you:

- Change the hourly rate?
- Reduce overheads and any other costs?
- Contract out if you have too much work?
- Recruit or dismiss staff easily?
- Vary overtime to deal with variations in workload? If you can sell more hours at £30 an hour, when the direct overtime cost is only £9 an hour, plus, say, £5 an hour for supervision (spread across two technicians), you make £16 extra net profit for each hour. A very cost effective way of operating. Or conversely, by increasing hours on that basis, you can reduce the cost per hour for other customers which you wish to retain or recruit. Those that try to control cost budgets by reducing overtime without consideration of the consequential effects can easily reduce income and profit.
- Add a sideline to absorb spare capacity - such as the firm that made wheelbarrows as a sideline in quiet periods?
- Increase marketing to generate sales. Do you have the structure in place to encourage more volume if needed? Does this increase the volume within the market place or are all your competitors chasing a limited or declining market?

And of course you should be aware of what the competition is up to, such as their rates, their costs, their market initiatives? Are you on top of the competition, ahead of it, or behind?

I mentioned earlier about the period over which income must exceed costs, i.e. to generate profits which is a natural requirement for any business. Some stakeholders or investors may be happy to look for long term, perhaps bigger and continuing, profits rather than immediate gains. But just as important as profit, even more so on some occasions, is cash flow. If the money in the bank is low and bills have to be paid, then selling more volume even if it costs more than normal, can at least bring in money to pay the bills. This does not have any effect on efficiency except maybe to make it more difficult to manage the normal operation properly. But increased short term cash flow may prevent the organisation from going into liquidation which is not efficient at all. A policy of continually working with short term critical cash flow however is not a good idea and longer term arrangements for increased working capital should be pursued by the business. Nothing however should stop your organisation from submitting invoices as soon as any work is done. Progressing outstanding payments is also a useful exercise which will normally be carried out by the credit control manager but operating managers need to know who is not paying as any further work, although profitable on paper, could be a complete loss of money if the customer is unable to pay and goes into liquidation.

THE RELATIONSHIP OF CHARGE VERSUS COST

What you charge will generally be what the customer is prepared to pay. What is the point of charging less if the customer wants to pay more providing of course he gets the service that he is expecting for that cost. **What you charge does not need to bear any relationship to what it costs you.** Even if you have to recover all your costs in an internal cross-charge system, the principles on which individual charges are allocated can be a matter of judgement.

Management accountancy allows you to allocate the total costs across units in whatever way is felt best for the operation and of course the market. How this is done will depend on many factors, such as the types of products or services being sold, the market into which they are being sold, and perhaps in some cases the inflexibility of existing finance and management accounting systems.

The positioning of your services within the market in relationship to cost and quality of service is part of the fundamental policy of the organisation and needs to be carried out in conjunction with your sales and marketing colleagues as directed by the organisation. It is no good deciding locally to provide a cheap, sorry, I mean budget service, if the sales team are trying to sell a quality service. Conversely there is no point in account managers trying to sell a service if both they and your team are not considering the same service standard.

If you are not operating in a niche or specialised market, a primary factor to cost reduction is volume. An increase in volume can often be managed by the same number of staff up to a certain level, increases in staff are normally at the lower salary level, or buildings can be utilised more fully, for instance by shift work. All of these can reduce the unit cost. But they depend very closely on work available.

Marginal costing refers to charges which just recover their expenditure. This is all part of the art and consequently there can be some misunderstanding. If you have recovered all your overheads from your "normal" customer base, paying the full rate or the agreed discount structure upon which you calculated your budget for the period, then any additional "contribution" above any actual costs will be further profit. However if the volume is below budget so that your normal work has not recovered all your direct costs and overheads, then marginal costing may not help.

It all comes back to the art of ensuring that enough work is sold or charged for over any period to recover your direct costs and any contribution to overheads that you need to make and provide some excess, the profit. How you charge depends on what your customers will pay.

Whichever way you charge, there is a risk. If volume reduces, first of all you lose profit; and that does not take very long. If volume increases, profit can increase quickly as you have already recovered your overheads unless this extra work is costed marginally. The sensitivity of volume to profit depends on the ratio of fixed costs to variable costs.

ACCOUNTABILITY

So you have worked out what you need to charge for a given volume. Your finance director is satisfied that over the period - say, a year - the projected income will more than cover your costs and contribute towards the company's overheads. As a budget holder there is one overriding principle which you need to consider. **You cannot be held responsible for costs or charges which you cannot control.** So if someone else within the organisation is dealing with the waste disposal contract, you need to be sure that such costs that will be imposed on you or that you are forced to accept are fair and truly reflect what you think you can achieve yourself, or less of course, especially if a large organisation uses its volume to achieve a good price for your small unit. Similarly, any budget performance must obviously exclude company overheads although you will be required to contribute to them. Otherwise your baseline will be increased.

RECONCILIATION

Any costs and charges need to be fully reconciled over the complete organisation. What you believe to be your costs and charges must be added to all the others to ensure that the total reconciles with the organisation's overall costs and charges. A simple principle which all accountants carry out in their sleep but one which is not always followed by others.

PERFORMANCE MONITORING AND BENCHMARKING

So how do you know at the end of all this whether you have cracked it?

Obviously one factor, perhaps the ultimate if you are providing services to real money-paying customers as opposed to internal cost centres, is profit. Indicators of profitability within particular industries are available from financial services companies which accountants and credit controllers regularly access. This will give a general indicator of profitability but is unlikely to be available until well after the end of the financial year when sufficient companies within an industry sector have filed their annual reports with Companies House. This may be too late to be of much use.

The hourly rate is a useful indicator based on number of productive hours sold divided by the total cost of providing that labour. If the cost of providing labour is more than you are charging - either the costs are too high or the hours sold are too low - then you may not last long unless the chargeout rate is increased. Do not include the cost of parts support staff in the labour rate if their costs can be recovered through the margin between cost of parts and sale of parts.

Utilisation of the productive staff should be checked by recording total staff hours available for work against actual hours worked. And finally, record the productive efficiency of hours sold against hours worked. These figures should be calculated weekly but may need to be reviewed over each four week period or using a moving four week period to iron out fluctuations due to work in progress or staff on holidays or training courses. A lot of this depends on what information is recorded. But a computerised workshop control system can record staff attendance times, job target (standard) times and job start and end times. This will enable the computer to quickly report the above indicators immediately after the end of the selected period.

These factors should be enough to give you a good idea of your overall performance. If benchmarking information is not available, you can monitor trends of your own unit's performance over the months and concentrate on the factors which you believe can be improved, such as personal productivity or better work loading.

Benchmarking is the comparison of performance indicators between different organisations or perhaps different cost centres within an organisation. This can be useful if comparisons are made of similar organisations and on the same basis although most of the time there will be factors that prevent direct comparisons. Benchmarking can give useful results if an independent organisation is involved such as a trade body, service organisations such as those that offer to monitor fuel and other commodity prices, or a vehicle manufacturer who offers a service to monitor its own dealers. However I know of no organisation that offers this type of service for vehicle repair activities to fleet operators or independent service garages.

AND FINALLY

I cannot guarantee that by following all these principles and ideas that you will achieve the perfect operation. But I have no doubt that they will help. And if they do, then I will have succeeded in my mission to share some of the lessons that I have learnt and developed over the years. You will not only be able to count the costs but also deliver a better, more cost effective and profitable service. What more could you want?

Transcript of the oral presentation
by Kevin French

Human Assets – A Technician's Point of View

KEVIN FRENCH
Motor Transport Technician, BT Group Motor Transport.

I'd like to talk briefly about a few of the initiatives that British Telecom (BT) have implemented to improve the workshop efficiency.

We've just started a new scheme of service schedules and standard times. At first a lot of the Technicians felt extremely paranoid that this would impact negatively on the repair times. But they soon realised that these initiatives were there to help them, to enable everyone to raise their standards within a reasonable repair time. At first you don't feel it's there for that, but you start to use these different things to improve the workshop efficiency and you soon realise that it's the way forward. We've always got to look and benchmark against other people.

Another thing that we've been doing is daily review meetings. Before we start work everyone gets together to discuss what the day's work is, so everyone can see what their goals are and what they're expected to do for the day. Therefore everyone in the workshop can effectively communicate with customers because it is clear what level of workload is achievable during the day. And I find that is an extremely good way of getting everyone involved in improving the workshop efficiency.

Similarly, we have a Quality Board that keeps an exact record of how much warranty remains on the vehicle you're servicing, and how it effects your budget. Which again reinforces to the mechanics what's expected of them, and they can see their goals.

Another initiative is participation in the Investing in People scheme, which standardises the training, goal setting for everyone in the workshop. I think perhaps that the ongoing standardised training is the most important aspect of striving to Improve Vehicle Workshop Efficiency. Because if you haven't got the training, you're never going to get there.

I'd just like conclude by saying that everybody in the work place can improve workshop efficiency so try to involve everybody with it from the tea boy to the highest manager. Thank you.